Contents

Picture credits: That lovely cover art is by Paul Dixon (pauldixonart.com). All the photos in the pictures chapter are © British Motor Industry Heritage Trust, references as follows; BMIHT Reference Number: Z-TowersPichestsrieder-1994, Code: A154851, BMIHT Reference Number: 00-NoelEdmondsRangeRover-1996 (1), Code: A154170, BMIHT Reference Number: 00-Rover800Production-1992, Code: A156123, BMIHT Reference Number: LJC-Rover75Production-03-1999, Code: A156139, BMIHT Reference Number: A154415 (smelting k series engine blocks), BMIHT Reference Number: 00-modellingAustinMontego-1980s, Code: A154931. All the pics in the chapter 37 are by the Author or his fellow automotive adventurers, I can't quite remember.

Don't sue me

A definition of the word 'story' is *an account of imaginary or real people and events told for entertainment*, and that's exactly what this book is; A story. It is largely, but not exclusively, based on some of my first-hand experiences and as such I have made changes to names, locations, products and actions for the sake of entertainment.

But it isn't a work of *total* fiction. Some characters are exactly as I remember them, others are inspired by people I met to some degree or other, none are totally dreamt up. There are facts with a sprinkling of twaddle, you can decide what you want to believe, my aim is to make you smile and think back to an era when the automotive industry was undergoing significant and often awkward change.

When I wrote the first draft of this book I was told that if I were to publish it as a work of fact then the weight of the corporate world would come crashing down upon me, and I don't want that. I have no axe to grind, no dislike of any company mentioned, and I'd happily buy and drive any of the cars mentioned here. Even Rover, who I've become strangely fond of long after their factories have gone. I'm absolutely certain that all companies mentioned make far better products today than when most of the events in this story happened, over twenty years ago.

The people in this story, indirectly, helped to pay my mortgage and feed my kids, and for that I am genuinely grateful. And besides - even the worst experiences with the looniest of them was character building. Even meeting Mr Tefal. So, this is just a story. Don't sue me.

Introduction

I knew this job was going to be entertaining when, on my first day in the office over twenty years ago, I accidentally insulted a colleague who then attacked me. As we rolled on the floor, me trying to apologise, him trying to staple my scrotum, the owner (an eccentric British inventor) walked in. He cheerily announced "*good morning, chaps!*" and stepped over us.

I loved that job; I loved the people and, as a young man, being given a role in sales meant the opportunity to travel and see some really interesting things in the automotive world. And earn a few quid for my young family. I am not an Engineer. Some might say I am not much of a writer, either.

This is simply a loose collection of stories based on my experiences of the various bodges and balls-ups I experienced (and sometimes caused) in my career selling quality inspection equipment to car and car component manufacturers in the Nineties. There's no logic in naming employers or companies or delving into the geekier details of what we were actually measuring. There are far better books on the important business of quality control and measurement than this one. They are probably quite dull though.

Today I do something else for a living but look back with fondness at the madness I encountered in quality control and on the shop floor amongst bodgers, skivers, union hard-heads and out-of-their-depth characters that helped make the job so much fun. Even with a scrotum covered in plasters.

Rover MG RV8, Oxford, UK

This wasn't my first job with the company. I had started as office boy, administrative dogsbody and general paperwork shuffler. My qualifications are pretty modest. I spent much of my education dossing about in classes held by largely uninspiring teachers who were constantly plotting strike action. I hated school, mucked about at college and after a spell swinging the lead at a small printing company who hit financial problems (not my fault), a large woodworking company who went bust (probably my fault) I winged an interview for a small but interesting measurement equipment manufacturer who shall remain nameless.

At work I found I liked talking (and listening) to customers, and was interested in how the instruments measured the way things were made. I hadn't really been into cars until that point. I drove a Vauxhall Chevette. I was asked to go and see a few customers. This quickly progressed to "*here's your company car, there's some demo kit in the boot, bugger off and don't come back until you've sold something*". I had to think on my feet. I liked that.

I remember making my first sale, at a plastics plant in Corby, and over-enthusiastically shaking the guy's hand about eighteen times when he agreed to buy my equipment. It felt good. I was prepared to put in the effort and thought nothing of driving 500 miles in a day at the whiff of something interesting. The first really big customer I dealt with was Rover.

Rover's Oxford factory was founded in the Twenties as Morris, a company that at one time made 51% of all cars sold in the UK. The Nineties were an interesting time for me to visit as they assembled Morris designed Rovers (MG RV8), Honda designed Rovers (600 and 800), Rover designed Rovers (75) and finally BMW designed MINIs. The factories, in Cowley, on the outskirts of Oxford, were well past their prime. British, Japanese and German production methods all overlapped; from a quality aspect it was a dish of roast beef, raw fish and sauerkraut. Served cold.

MG and Rover were then the same company, but the MG RV8 model was a throwback to when MG was a separate brand, with their own models. I never actually got to see it in production. This was a model originally made by MG from 1962 to 1980 as the MGB Roadster and, for whatever reason, Rover decided to exhume it for another production run from 1993 to 1995. Drum brakes, live rear axle, leaf springs and all.

The bodies were made in Witney and the cars completed in a wet shed in the corner of the Cowley site. I only managed one meaningful visit there ever and the longest conversation I had with their quality control department was apologising for leaving a door slightly ajar to an angry inspector sat on a stack of cardboard boxes of components, who was trying to read The Sun, undisturbed.

It was an interesting introduction to Rover, for me, because I had the feeling that they weren't very good at quality control but couldn't find anyone there who seemed to care about it. In the MG RV8 production department I never did. Rover made c.2000 of these outdated oddities and many were sold to nostalgia hungry Japanese anglophiles. Many of these ended up back in the UK when, presumably, oriental owners realised that even new RV8s drove like they'd been Fukushima'd. But I had got my foot in the door of Rover and was loving the job.

Rover 800, Oxford, UK

You might remember the popular Rover 800 TV advert, it was in German with English subtitles. Two hard-faced Germans are driving serenely along in one of Cowley's executive saloons. Hans compliments Fritz on his car, who smugly replies "*Britischer Arkitekt*". It's a great advert, but I know of no British designer with the name *上原 繁*. The 800 is largely a Honda Legend, a car designed in Japan by a chap called *上原 繁*.

A deal struck between Honda and Rover meant that Cowley actually produced a small number of Legends in the eighties, but they were so badly slung together that Honda refused to accept delivery of some of them and they mostly ended up as transport hacks around the site, being unfit for the open road. There were one of two of these shonky freaks still chugging about in the mid-Nineties when I was involved with a quality improvement program for the facelifted Rover 800. I supplied quite a bit of equipment in the end.

Many car builders use SPC (Statistical Process Control) for measuring production consistency. One common statistic calculated is called C_{pk}. My equipment would make a number of measurements then automatically compute this. It was generally accepted in the automotive industry at that time that a C_{pk} value of 1.33 was acceptable, anything less was not. I apologise for the geekery here, but it is relevant.

Engineers would make measurements on a daily basis and any process with a C_{pk} of less than 1.33 meant they ought to take action to improve the production process, or (if it was really bad), stop the line. In addition to this, our equipment's display gave a red or green light with each measurement. You don't have to be a quality expert to work out what they meant. One particular inspector at Cowley refused to believe the display and insisted that a red light just meant the battery needed charging. He had worked there for ten years and hadn't seen a green light before.

It was a hopeless situation. Results varied wildly, no two 800s were built with the same degree of accuracy. Some were OK. Others needed a lot of rectification. Rover really wanted to sell the 800 in America but buyers there were not impressed. One common quality complaint from the colonies was that the 800's leather seats, when exposed to sunlight, went green. I worked quite closely with one engineer, who we shall call Malcolm, who I thought was really interested in improving results at the factory.

Malcolm had an accent that made him sound like a West Country farmer. He once showed me the Rover 800 Coupe production area. There were holes in the roof, puddles on the floor and a flowering buddleia bush growing in a corner. An advert for the Rover 800 Coupe boasted about them being 80% hand made but it was a turgid time for them - they just couldn't make them right.

I gave Malcolm a lift home once, after his company car (a Rover) refused to start. I dropped him off and he invited me in for a cuppa. I shouldn't have accepted, but did. On a wall was a huge poster of a scantily-dressed boyband. Seeing that I was a bit bemused he explained "My girlfriend likes 'em! She's still at school." To this day I don't know if he was talking about a schoolteacher or a schoolgirl.

Traceability, RTFM

A product, such as a car, is only as good as its components and the way it was designed and built, and to ensure it's made to the right quality many things need to be measured. Your measurement is only as good as the instrument you use, and that is only as good as its calibration. Bear with me, I'm getting there. The calibration needs to follow a procedure using very precise known quantities, such as a reference weight or length, for example.

In various laboratories around the world, such as the National Physics Laboratory in Middlesex, UK, they have shrines to lumps of stuff used as references for these measurements. Even the gravity in their rooms is measured. This is traceability, from the shop floor all the way back to this lump of something. Everything measured needs a bit of paper enabling an auditor to follow this trail accurately.

This was a weakness in so many companies I dealt with. They'd have some battered old test instrument and point to a PAT test sticker and claim it meant the readings were accurate. It's rather like dimwits thinking an MOT is the same as a service for a car. We had calls from customers asking why their measurements were inaccurate or, in the case of Rover people, claiming "*your thing is knackered*", when they'd last had their kit properly calibrated years ago. So we installed a bit of software that warned the user if the kit was out of calibration before they used it, which worked for everyone but Rover people who would call and say, "*your thing is knackered, it's giving me an error message or summat*".

The boss had had enough of Rover and their inability to RTFM (Read The Flippin' Manual). On one complex bit of kit, where the operator had to pause between measurements, Rover QA operators would jab away at the display impatiently, causing it to crash. The boss changed the '*please wait*' error message to read '*stop f*****g pressing buttons for a minute you bloody communist!*'

We didn't get many requests for support calls after that.

Rover 600 and the Honda Accord, Oxford and Swindon, UK

Rover's 600 and Honda's Euro-spec 4th generation Accord are almost the same car. The Accord was assembled in Swindon, the 600 up the road at Cowley. Both companies were customers and I used to shuttle between the two in my company car; A Honda Accord.

There was an occasion at Swindon when we once found a small spelling mistake on a purchase order. That's the only time I ever experienced anything out of the ordinary there. They had no quality issues. My own Accord was super-reliable. It did 60k miles a year for 2 years and, with a 6k mile service interval, got serviced nearly once a month. Which was the only time it ever got cleaned. I had done 25k miles in it before I even realised it needed a service and every gear change was bounced off the rev limiter. Nothing went wrong. Ever. A perfectly reliable car.

Yet a few miles up the road *almost* the same car was being built by Rover to *not quite* the same standards. The guys in quality of both companies used to talk and on paper their results were comparable. Honda never knew why their slapdash Cowley mates made rough cars but got away with it. Well, I knew why. Rover published their QA stats but simply edited out the worst 1/3 of the results. It was like showing poor homework marks to your parents and using a biro to carefully craft the grade E into a B. Sorry, Mum and Dad, if you're reading this. It was an E for me, too.

Rover 75, Oxford, UK

Yes, Rover again. It's easy to knock them. There's not one single thing that killed this once great company; it must have been a multitude of bad decisions over a prolonged period that caused their eventual demise.

These stories might make them sound worse than they were, but as you'll see from the number of chapters in this book, nearly all factories I visited made the odd cock-up. Rover seemingly just made lots, all at once, over a number of years. I would never have bought one in the Nineties, but today I would quite happily have a 75 Tourer on my driveway.

The 75 was initially made at Cowley to a Rover design. What looks like a transmission tunnel is often mistaken for the rear-wheel drive underpinning of a BMW 3 Series but it's a coincidence - the 75 is a thoroughly British car and a good one too.

When BMW finally split Rover off and left it to the vultures, Cowley's 75 equipment was sent up to Longbridge and not *all* of it was sabotaged by disgruntled employees. There was a great documentary which summed up Rover's malaise to me. You can probably still find it online. It is called "When BMW met Rover" and was a fly-on-the-wall thing filmed during the time that BMW took over.

One scene showed the young engineers at lunchtime in the canteen at (I think) Longbridge. The Brits and Germans were supposed to mix but didn't seem to. At the German table was earnest talk of tolerances. At the British table the silence was broken by a bored Brummie voice; "*what's for puddin'?*"

Editors can be cruel, but that sort of summed up the difference in attitudes to me; when Cowley started making MINIs things got even more interesting…

An idiots guide to production terminology

The world of car production has its own language, often wrapped up in natty acronyms, designed to confuse normal people. Here's an idiots guide to some of the more common terminology:

8D – A problem solving method, originating from Ford. The 'D' stands for disciplines; the 8 of which starts with plan and ends with congratulate, assuming you ever get to the bottom of the cock-up.

APQP – Advanced Product Quality Planning is a snappy acronym invented by the big three (Ford, GM and that other one), the aim of which is "to produce a product quality plan which will support development of a product or service that will satisfy the customer". I did warn you that the geeky bits were dull, didn't I?

Body in White – The body of a car before painting or assembly has taken place. There was a Jaguar version of this phrase; on visiting their Browns Lane plant in Coventry I once witnessed a batch of unpainted XJ-S bodies sat outside in the rain and was told they were 'body in shite'.

BOM – Bill of Materials. A shopping list of bits needed to build your vehicle. Like a new recipe, there's often something important missing that you don't spot until serving up.

DFSS – Design For Six Sigma. This is a set of methodology related to Six Sigma. It is not to be confused with DFS, the *buy now offer ends Monday* sofa warehouse chain situated in Darley Dale, Measham, Droitwich and Grantham.

FEA – Finite Element Analysis. The use of a very expensive bit of software that requires regular and cripplingly expensive updates to produce molecular level analysis of products. In pretty colours.

FMEA - Failure Modes Effects Analysis. Originally designed by the military FMEA is a complex technique for failure analysis. FMEA analysis can relate to design, function or process.

ISO – International Organization of Standardization. There is an ISO standard for almost everything. The German arm of ISO is DIN, *Deutsches Institut für Normung*, with overly-officious standards probably written by the kind of people my Grandad used to enjoy shooting at.

ISIR – Initial Sample Inspection Report. A VW and BMW flavoured version of PPAP.

MRP – Materials Requirements Planning. This is a planning, scheduling and inventory system, first used by Black & Decker.

MSA – Measurement Systems Analysis. This is an experiment designed to identify the components of variation of a measurement. Often conducted by humourless chaps in lab coats who won't let you touch their stuff, no matter how interesting it looks.

NVH – Noise Vibration and Harshness. A very important subject in the automotive world. Companies benchmark cars they have in development against class leaders. NVH measurements are made using sophisticated equipment strapped to test vehicles driving over cobbled test tracks, or sometimes some poor unfortunate simply gets strapped into a car boot with a microphone and driven around for hours listening for squeaks.

Pareto Analysis – A bar graph that looks at all the reported problems and helps identify which problem is most severe and should be addressed first. Named after 1991 world snooker champion, John Parrot, perhaps.

PPAP – Production Part Approval Process. A 19 element process designed to ensure suppliers can reliably meet the requirements of supplying a part that meets the car manufacturer's expectations. This is related to the car salesman's acronym TADTS, standing for *they all do that, sir*. One reason they don't change shonky bits in production is because PPAPs are so very expensive to produce.

R@R – Run At Rate. A calculation to determine the output rate of a production line. Probably also something to do with cricket.

SCM – Supply Chain Management. They'll try to have you believe, using further acronyms, that there's more to SCM than purchasing. But there's not. It's just purchasing. To any automotive SCM types reading this; you might have had a fairer explanation of your work here had your fellow night creatures not sucked every last drop of profit from everything I ever did that they got their fangs into.

Six Sigma - The six sigma goal for manufacturing is to drive quality to less than four defects per million parts built. The Wall Street Journal once estimated that nearly 60% of all corporate Six Sigma initiatives fail to yield the desired results. Probably because it's far easier to say to a workforce "less than four defects per million, please, chaps" than it is to actually roll your sleeves up and achieve it.

SPC – Statistical Process Control. A method of quality control using statistical methods. Unless you occasionally used to ignore the worst of the results to make your graphs look prettier than they should (ahem: Rover).

SQE – Supplier Quality Engineer. The person who, when your product is cack, turns up and ensures you put it right. Japanese manufacturers employ roving teams who will reside in a supplier's factory until any quality issues are resolved to their satisfaction. These people are ruthless and do not eat, sleep or blink.

TQM – Total Quality Management. A system to implement progressive improvements in quality.

TS 16949 – This is the document that defines the quality management systems and requirements for all products within the automotive engineering and manufacturing world. You can't do business without it. Hooky Chinese outfits, such as the 'Love you Longtime Lucky Ballbearing Company Ltd', will have a photocopy of someone else's TS 16949 certificate.

MINI, Oxford, UK

No, my caps lock button isn't stuck. Apparently the new (BMW) MINI must be written in this shouty fashion to distinguish it from the old (Rover) Mini.

I used to occasionally bump into an old engineer at Cowley. If he were in government he would have been a minister without portfolio. He sort of wandered around and knew everything, without actually doing anything. He told me that when he started at Morris they made many different models, and even bodies for Rolls Royce at one stage, but now BMW were turning it into a factory to just make MINIs. He was retiring because this made him sad, and "*you can't trust Germans*". I think that was rather misguided because BMW really turned the place around, but not before a few mishaps.

I was fortunate enough to be given a tour of the new MINI line during pre-production. It was a totally new facility and all top-secret. No cameras allowed, any metal had to be taped up to prevent scratches, paperwork signed and all very Teutonic. I was given a tour by a particularly officious German HSE Manager, new to the company.

As we walked the line the world's first few MINIs were slowly but carefully being assembled, and some of my equipment was being used. It felt like history in the making but this man was a bore. He wanted to talk about safety, safety, safety. As we got level with one operation his safety drone was drowned out by a scream and lots of shouting. A man had his finger caught up to the knuckle in a machine. I didn't see Herr HSE again after that.

Rover at Cowley was no more and, a few years later, the whole company finally went under. I now look at a sea of 2-litre German diesel saloons on the roads and long to see a Rover 620ti, or an 800 Coupe in British Racing Green. Even a Montego Countryman diesel would get me interested. There. I've said it. My dirty secret is out.

Volvo, Gothenburg, Sweden

Volvos are built at various factories around the world. There is production at Gothenburg in Sweden, design and quality management too. The motto of the Gothenburg factory is “Increased capacity – for ever-higher quality” and frankly they were always far smarter than me. They even spoke better English. They made such good cars that there was little reason to visit. But I did anyway, because their canteen was so good.

Anyone with a visitors badge could dine for free and whereas British factory canteens usually offered grey slop and gravy, Volvo had salmon, salads and little cheesecakes with fresh blueberries on. Twenty years on, typing this, I am salivating at the very thought of those cheesecakes. All this wonderful nosh on the flash of a visitors badge. A visitors badge I stole on an early visit to Volvo and brandished in return for a free meal anytime I was even remotely in the vicinity of Sweden.

Skoda, Mlada Boleslav, Czech Republic

By now I really had the travel bug. Production problems were quite similar at most of the factories I had visited - and similarly resolved - and as my confidence grew I looked to jump on any flight heading somewhere interesting in order to get a few more stamps in my passport. My next stop was Skoda. My Grandad couldn't remember which side the Czechs had been on in the war and tried to lend me a large, wooden-handled screwdriver "just in case, lad", before falling asleep.

VW had owned Skoda for a few years already and were beginning to slowly exert their managerial style, but we had a local agent who was well connected, so when I visited the Czech Republic I thought business would be as straightforward as it was in the car factories I dealt with elsewhere. Then, Skoda was rapidly recovering from communism and had factories in three towns, each with a name that looked like a bad hand at Scrabble. Their HQ was (and still is) in a town called Mlada Boleslav, named after Boleslaus II the Pious. The Pious bit was rather apt as Skoda were seen to be the budget end of the VW brand.

I had been brought up on the classic salesman's training of 'features and benefits' and, combined with a strong product and application knowledge, I was sure that Skoda would snap up my QA inspection equipment and I'd earn a few bob commission. I flew to Prague with my box of tricks and drove to the town of Mlada Boleslav [triple word, double letter score = 187 points] where the main factory was.

The Czechs, when I first arrived at Skoda, were downright rude. I don't speak Czech. My German is pretty good (don't tell Grandad) and I couldn't understand why I got evils from everyone I met. Surely as a German run business they'd speak it? My "*Guten tag*" was met with a look of hate. They actually thought I *was* German, and from VW HQ. An apologetic "*Ich bin Englisch!*" worked wonders and after a forthright discussion about Nazis I had a tour of the factory.

At the time Skoda made the Felicia, which was a freshened up Favorit and very much Skoda's own car. It had been designed before VW were involved and the production line was like something from the '70s. Or Peugeot's 306 line at Ryton in Coventry in the '90s. Anyway, nearly everything was done by hand and the model was soon to be discontinued. Operators in most factories performed about three to six operations per workstation. Here, men with spanners and moustaches did one operation and glared at me as if I were Himmler himself.

I looked at a particular component assembly and asked Mr Moustache what the tolerance was. I got a blank look, so tried to explain myself better; "Plus, minus… ?" He shrugged and said "*plus, minus... autobus*". Give or take a bus. I appreciate this this isn't a comprehensive assessment of the state of quality control at Skoda, but it gave a glimpse of what things might have been like pre-VW.

We were shuttling between departments and getting to see much of the Czech countryside. Beautiful buildings in that mustard yellow plaster around cobbled squares where locals smoked, stared and smoked some more. I was told of the arrogance of some Germans visiting. Local Police were happy to pull them over for a bribe; the Germans were (comparatively) too rich and too busy to stop, so in answer to the flashing lights of a Police car they'd simply post a banknote out of the driver's window without bothering to stop.

Our agent was so confident we were going to win an order that he decided to celebrate. I insisted on paying in the hope I could keep a lid on his enthusiasm. He booked the best restaurant in town. I had been warned, before this trip, that the three worst things in the Czech Republic are breakfast, lunch and dinner. We went somewhere really fancy. Our man had quite a few drinks and decided to show he was Mr Big.

Red wine was served, but returned because "it wasn't room temperature enough". The bottle that was brought back to us was so hot you couldn't touch it. A starter of steak tartare was returned with insults, he insisted on having it "Cooked! Medium to well done!" Tomorrow's big meeting was going to be a tough one.

After an early start we arrived at the brand-new Octavia (type 1U) production line, the main reason for my visit. I got my kit out in the QA lab and explained that we were going to test some suspension components and compare the results with the specification, as written by VW. I expected a mess.

The Felicia was a simple machine and with sufficient (cheap) labour on the line everything could be double checked and corrected as they made it. The Octavia's production line was brand new, more automated; the (German) design more complex and if we were going to find something even slightly wonky it would be in the suspension.

I could usually find something out of tolerance using my equipment, and that was usually enough for the customer's QA department justifying having to buy it. The agent and Mr Inspector whispered in Czech as I checked about ten parts. Every single one was spot on, they were even better than results than I'd seen at Wolfsburg. It turns out the men with spanners and moustaches actually really knew their stuff.

I apologised to our agent and explained that the car was perfect and they really had no need for our equipment. We'll buy it anyway, said the Inspector, to my surprise. The Buyer, the Engineer, the Manager, we met them all. They all signed it off on the spot and we won a nice order somehow. As Steve Coogan's salesman, Gareth Cheeseman, would say; "Back to the hotel for a w*nk!"

I was taken to a different hotel in the sticks for the evening. "A fancy place", it was owned by a friend of our agent. He told me it was a kind of health spa. It was "very expensive", the equivalent of £7 a night (cash only). It looked like a castle. Few windows or lights. Or guests. It had gloomy ornamental gardens and I went for a walk in the grounds. I was accosted by a chirpy teenage lad with a Californian accent. He was delighted to have someone to talk to. So was I. His parents had sent him here to help him get over some "addiction issues". He was a wild-eyed freak. I couldn't shake him off and so went to bed.

The receptionist (a beautiful lady in a long white coat) escorted me to my room. How come so many Czech women are so beautiful, yet their men often look like potatoes on legs? Surely, when they make children, the genes would mix and they should eventually end up with an average looking population? I have never been able to understand this. Anyway, the 'hotel' room had a white tiled floor and walls and ceiling. And there was no handle on the inside of the door. She locked me in.

In the morning I was collected from what I now realise was an asylum, by our agent, in his Octavia. I recall thinking; this is a *really* decent car. I asked him how we'd sold such an expensive piece of inspection equipment when Skoda had no need for it. I've removed the rest of this sentence on legal advice. Nothing to read here. Move along, please.

The Czech guys at Skoda we met were great engineers, but Skoda isn't the only Czech car company. Tatra was an innovative manufacturer founded in 1850 and there's a bit of anecdotal information online to suggest Porsche pinched the design of their V570 car when they designed the Beetle.

The 1990s marked the end of Tatra as a car manufacturer. Their chunky T700 powered by an air-cooled, rear-mounted V8 was finally outdated. I thought it a real pity that their engineering brilliance wasn't taken under the wing of one of the automotive industry's bigger players at the time. Search YouTube for a beautiful, retro promotional film that they made called "Tatra – Happy Journey". I'd had quite a happy journey in the Czech Republic myself. Next: the Netherlands.

Volvo, Born, The Netherlands

Volvo's factory in The Netherlands was actually a company called Nedcar who built the S40 and V40 on Volvo's behalf, and the very similar Carisma for Mitsubishi. These guys had mentioned having occasional quality issues and I was invited to take a look and see how I might help.

On arrival I was given one of those Dutch 'sit up and beg' type pushbikes with a basket on the front. There was an undeniable fug of pot about the place, as I recall. The quality department was at the far end of the site so I enjoyed a long indoor bike ride behind a dope-smoking man in dungarees and I cycled along downwind of him for some distance. He got to the office we needed and stopped abruptly. I wasn't really paying attention and my reactions were a bit blunt. At that precise moment I found that there were no brake levers on the handlebars, but back-pedal brakes. I made quite an entrance.

The nearest first-aider tended to my grazed hands but I bravely declined assistance for my heavily bruised genitals. I don't remember much else about Nedcar apart from a hundred production workers laughing at me.

Reader's rants

My stories are mostly told from the perspective of me, as a salesman, working for a supplier to the car factories. When sharing these stories I love to hear other people's experiences in return, particularly those with a different perspective to mine. As I was writing this book a few people got in touch to share their rants with me about brands I didn't know much about, and I've included them here.

Let's start with someone who suffered life as a service manager for a major car dealership, who goes a bit BoyZone on us, as follows:

"Ronan Keating sung about saying it best, when you say nothing at all. I have "nothing" to say about Pendragon PLC. Signed - An Escaped Service Manager (or "Leader" as they like to call them).

Ouch! I had to remove the rest of his rant on legal advice, which was a pity, as it contained a wonderfully colorful phrase about bottoms. And from anger to a beautiful description of someone's rubbish first car. Eric Rood, from the colonies, speaks from bitter personal experience:

"Mercury Topaz: The automotive equivalent of soiled underwear you unsuccessfully tried to flush down the school toilet in fourth grade."

I had to google the Mercury Topaz. I wish I hadn't.

It's not all bad vibes; rants can be used for good as well as evil. Jeff misses a Swedish brand not covered elsewhere in this book.

"Safe, economical, stylish, comfortable, well-built, turbos. Why exactly is Saab now dead?"

I think I could answer that but the brand-swallowing monster of GM might swallow me too. Saab are now NEVS, or something, making electric cars. Apologies for the apathy, I can't get excited about new cars.

From the other side of the showroom floor comes another perspective. Now, I can relate to this. I had a friend who was high up at some Peugeot dealership in the Midlands who once explained that his customers just weren't 'car people'. They were '*cheap tax, no deposit, do-you-do finance, innit mate*' sort of people.

AH doesn't sell Peugeots, I don't think, but he clearly knows what matters to his customers and it's not laser-straight shut lines:

"As a car salesman I appreciate 'quality control', I also realise it's completely wasted on the people who actually buy cars..." – AH

Baz (who makes an appearance in chapter 37 of this book) asks;

"How does the board of Ssangyong look at their designs and think 'that's a job well done, green light for production'?!"

Baz, you can blame a fellow Brit for much of it. The Ssangyong Rodius was designed by Ken Greenley, formerly of the Royal College of Art in London. Apparently the design brief was '*capture the essence of a luxury yacht*'. Their equally repulsive Rexton was styled by Giorgietto Giugiaro, he who did the fabulous Lotus Esprit, gorgeous ISO Grifo and iconic MK1 VW Golf. It seems even the greatest designers have off days.

And finally to Danny Phillips, who sums up his Land Rover ownership, thus:

I love Land Rovers, all of them, but could they please decide what system to use with their nuts and bolts. Metric, Imperial, Whitworth, Martian, just use one on the cars, not all of them on every car.

Thanks for these contributions. Let's get back to the factory floor, shall we?

Nissan, Sunderland, Mackem-land*

I never really experienced any quality control calamities at Nissan. Built at the outset as a Nissan plant to Japanese standards, and with a new and willing workforce they knew how to measure, manage and control their production to the highest standards. My visits were profitable but unremarkable. Workers there valued their jobs but every now and then you'd see a chink of anarchy as someone performing the same repetitive task over and over would go a bit doolally.

Many components for assembly were delivered on a 'just in time' basis; no stock sitting unused cluttering up the line; suppliers often being paid when the component was fitted to the car on the line. This requires an obsessive on-time delivery regime. If a component arrives even seconds late the line stops and lots of money is lost. The Japanese way of thinking is that you can't trust a man to deliver these parts in time. A machine should do it. And so, in one section, on an hourly basis (not a second too soon, or too late) a robotised trolley would arrive with just the right amount of parts on it. The worker lacked human contact on this section. It was a lonely and repetitive job. It would have been nice to have these bits delivered by a person.

The robot had a flashing light for safety. Which was annoying. If *your* peripheral vision was subjected to a flashing light on the hour, you too, might start to develop a deep hatred of the robot. The worker was smart. He realised that just moaning would make no difference to the po-faced Japanese management. So he raised it as a health and safety issue. This thing was silent, someone might not see it's flashing light and get run over he said, requesting that the machine was replaced by a person (who would also, he hoped, help alleviate his boredom).

Management had one of those stand-up meetings where decisions are made quickly but no-one gets tea and biscuits. Subsequently the robot was fitted with a warning sound so people would hear it coming, but took the workers mental state into account (so they said) and made this sound a short and warbly tune. I don't know the tune. I heard an ice-cream van play it once, and heard it myself when I was making a routine visit, but the robot delivery trolley vanished soon afterwards.

The worker, mind eventually scrambled by lights and inane plinky-plonk electronic torture, took revenge. The robot followed a pre-defined route, the shortest route possible, by detecting and following a metal strip fitted to the floor. The worker brought in a roll of metal tape and laid a new path for the robot to follow. It was last seen heading Southbound towards the A19.

**Sunderland was once famous for shipbuilding by day and theft by night. Locals are known as Mackems, from 'make 'em', so it stands to reason that those out stealing at night, taking 'em, were 'Tackems'.*

The man who stapled my scrotum

I was not alone in my work. Although we were a small company, we had some quite complex products and employed some electro-mechanical and software engineers to develop products, support customers and bail out people (like me) who's very best button-pressing didn't always yield the right test results.

The chap who stapled my scrotum on my first day at work, mentioned back at the start of this book, quickly became a good friend and useful chap to have around. We had a little lab where, on quiet days, we would tinker with equipment and learn how to get the best measurements with it. We'll call this chap Pedro. He had an interesting background; South American and Irish parentage, an upbringing in a rough corner of Newcastle and a career in the RAF. He was a handful.

I think he was bored a lot and as a result would do things like make me stand up a ladder then pass me a live 240V cable, with the wires carefully stripped to ensure I conducted as much pain as possible. On getting my first company laptop he made sure I had internet access to the most arcane bongo-sites before I'd even mastered MS Word. But when something went belly-up he was a great guy to have around. He told some outrageous stories.

He was obsessed with how materials performed under stress. This was related to the work we did. He claimed that his Father once built a bicycle out of papier mâché using rolled up newspapers to make the frame. He didn't last long before getting up the noses of management. He had a bit of a class-warrior thing going on and anyone he perceived as being in authority was in for a hard time. Fortunately, this wasn't me.

No-one was really sure how/why he'd left the RAF but I suspect it was related to a story he told me once, something I did get partially verified by someone in the know.

Pedro had once worked on the RAF's flight simulators. They didn't have the software-based systems of today; they had massive maps of Europe populated with tiny models of houses and bridges, in a huge hanger. One map might cover half of Belgium, for example. Over this would 'fly' a camera attached to a very long arm, which would respond to the pilots inputs. The Pilot sat in an adjacent room in a mock cockpit, looking at a monitor showing the view from the camera. Houses were 3mm tall, rivers a centimeter wide and everything was perfectly detailed. If the pilot flew off the edge of the map the arm lifted and placed the camera in a box of cotton wool to simulate clouds while they switched to another map.

Pedro didn't like Officers. He placed two large spiders in the cotton wool box. After hours of stressful flight simulation over the Low Countries, Officer Carruthers flew off the map and into the clouds to be met by spiders the size of a skyscrapers. Pedro used to enjoy telling this story and mimicking the Pilot's muffled screams.

This is the guy who, ultimately, made sure car manufacturer's measurements were correct. That your engine wouldn't go bang. It wasn't a surprise when he left the company but I miss him to this day. He flew to Korea once, at very short notice, fixed a broken bit of control software at Daewoo and rang me at 3AM, knowing I like animals, to say he was eating dog.

Toyota, Burnaston, UK

We had a competitor in the UK. Not a big one, but a company who were a thorn in our side. The rep was a gobby, barrel-chested Yorkshireman in a double breasted blazer with gold buttons. Let's call him Derek, for that was his name, and as he's now dead he won't mind me using it. He was a nuisance for us, and Toyota engineers alike, as he visited to check equipment and snoop around the production line looking for more business.

You can't just wander around a Japanese car factory without a chaperone, or paperwork. Workers have scant seconds to perform their tasks and don't have time for interruptions. Derek tried everything to ingratiate himself with people at Toyota in Burnaston including buying a new Avensis, made at the same plant. "*Come and have a look*" he said, showing a crass lack of understanding to a senior engineer. Why would a Toyota employee, trying to manage a production line making Avensiseses, want to go and look at a sales rep's own Avensis, in the carpark?

Derek had it coming. On his next visit he returned to the visitors carpark to find his car had vanished. It had been moved, said security, to another compound, handing him his own registration plates which had been removed. The shipping compound was packed with hundreds of finished Avensis, identical to his, parked hard nose to tail.

A Tractor Factory, Somewhere

I'm not going to name this place. I don't think it would be fair on the people involved and because it involves a fatality. It was a heavily unionised factory that made tractors somewhere in the world.

It was a long way from home and I soon learned to call ahead of an appointment to ensure they weren't taking industrial action before I departed. Sometimes my contact would agree a meeting and I'd set off, only to find that in the time taken to travel there someone had looked at someone else's spanner, or something, and they'd all walked out on some unofficial strike.

It reminds me of an old joke told about two famous union heavyweights. Scargill and Robinson were walking in the park one lovely spring day. Scargill said "I see the daffodils are out" and Robinson replied "official or unofficial?" There was always a reason to strike at the tractor factory.

It was, I recall, a horrible place. Production was done, due to the nature of the product, at a more pedestrian pace than the car factories. Chassis and bits were dragged from workstation to workstation, meaning departments became more enclosed, defensive. The site had a huge production hall, with lots of little offices dotted around, and to my untrained eye the diagram of production flow looked like a bucket of wrestling snakes. Shirking workers liked this. It made it easier to bamboozle management. But what a stressful place. Shop stewards ruled.

I used to visit three engineers, let's call them Tom, Dick and Harry. These three were as thick as thieves. They'd been school friends and had always worked at the plant. It took some time to win them over but their department became a good customer for my QA kit, even if they scarcely used it. I used to enjoy their angry tales of rebellion, and listen respectfully when they raged about people being fired 'just' for what amounted to pilfering or skiving. They took their breaks at the same time, sat in the same seats, and ate the same food every day. They worked to rule and management were on their case.

I visited one miserable winter's day to find that Tom was absent. I was then told a tale which you are free to disbelieve. These three Engineers had been given a whopping-great bollocking for something by management. It made Tom feel ill. Very ill. Later that day he stood up, clutched his chest, and fell down in their office doorway dead from a heart attack. I was told this soon after it had happened by Dick and Harry, long-standing mates of poor deceased Tom.

Someone senior had called Tom's poor wife and told her the sad news. At this point, telling me the story, Dick got quite agitated. "*F*cking management lied*", he said. "*They told her Tom didn't suffer, that he died instantly. I told her the truth. She deserved the truth. He took ages to die. He fell over at 10.50 and was still on the floor breathing at 11.00. I remember the time exactly because it's when I went for my break. When we got back, he was dead*".

I digested this info, knowing the office layout, and knowing the bond these men supposedly had. Tom had fallen in their doorway. Dick and Harry left him (with a first aider, presumably) to go on their break. To do this they would have had to step over his body to leave. And they complained about a white lie management had told to console his poor wife!

Until this time I had always enjoyed the tussles between unions and management, as a supplier, a neutral, I found it quite funny at times. This was just horrible.

Component factories

I had a spell of visiting lots of tier one suppliers. These are the companies making thinks like brakes, seats, plastic assemblies and other large parts for the car manufacturers. These companies varied; Some were well run and profitable, others were shambolic and always in trouble with their customer. The latter were the most fun to visit.

Some were ex-British Leyland plant spin offs, still making bits for Rover. The quality manager of one such ex-BL factory was a tough nut. They made fuel tanks which are essentially large plastic containers filled with a kind of plastic wool to help prevent the tank from bursting if crushed in an accident. The thinner they made the plastic, the more profit they made, but plastic can be a very variable material to try and measure. It was just so inconsistent and they were always cutting corners to save material costs.

The QA manager had glasses as thick as a Bo-Selector character. No matter what the result of my highly accurate digital test equipment he would always feel the component with his hands, squinting at it, bending it. I asked what he did before he joined this factory. He showed me a photograph of himself, as a referee, sending off a youthful Paul Gascoigne. He wasn't wearing his bottle-bottom glasses and Gazza was clearly upset. They bought the equipment but I got the impression that the ultimate test of quality was his gnarly-handed groping of parts.

Another plant made brake pedals for Rover. These are simple parts, a bent bit of steel with a couple of welded fixings, painted and finished. But brake pedals need to be *exactly* the right shape. It's a lever, and needs to function perfectly for obvious reasons. They were making such a hash of making this part that a QA manager at Rover instructed me to visit, and they reluctantly received me.

Like other old component factories the air was thick with the smell of cutting fluid, wooden floors ingrained with decades of oil, ancient machines chugging out parts from worn-out tooling. Their final inspection tool, the thing that checked every single brake pedal destined for Rover, was reluctantly shown to me. It was a go/no-go gauge made from an old piece of wood. They pressed every pedal they made, by hand, against this bit of wood - and as long as it had *roughly* the same contour the part was OK'd. This QA 'instrument' looked like one of those bits of chipboard found trampled on the floor in IKEA's Bargain Corner. "*This can't be acceptable*" I told them, "*Where did you get it from?!*" The QA manager turned it over. On the back in marker pen was "Do not remove! Property of Longbridge QA department".

I used to make many trips to South Wales. Traffic was light, the scenery was good, and I used to enjoy charging around knocking on the doors of interesting-looking factories. Today we have Google and telesales. Then it was all legwork and cold calls and I really enjoyed it, but I ended up in some weird places in the pursuit of business.

I recall a large industrial estate somewhere near Swansea. Every single factory was boarded up. I drove round and round and found a roadside caravan selling tea, coffee and burgers. These places were often a good source of information. The burger-flippers would happily tell you who the big local employers were and other titbits of info on local industry. I'd buy a cuppa and ask a few questions. This caravan must have done a roaring trade when the factories were open, but it was pretty grim now.

The caravan was a regular touring caravan, with one window removed and a rudimentary shelf screwed to it. The legs weren't down properly so it was leaning to one side. The two people inside were most peculiar. I would guess they were brother and sister, but each could have been either gender, or transgender; round shouldered, pale flabby faces and matching bumfluff moustaches.

I asked for tea. One of them boiled a kettle and poured hot water into a Styrofoam cup. "*wanmilkinert?*" I had no idea. They held up a carton of UHT milk and repeated the question. Ah! Milk! Yes please. "*wanshugarrinnert?*". One lump, please. I couldn't ask these poor pair anything about local factories, but their tea was nice and only cost 15p.

At that time I think some of the anarchy I experienced in these factories was starting to rub off on me a bit. At a brake factory in Wales I was kept waiting for ages in reception once. I sat next to a welcome board with stick-on plastic letters. 'Welcome - Glenys Kinnock', said the sign. The local MP was visiting later that day, the press were coming too. I wonder if they got a photograph of the welcome board after I'd found the box of letters and idly made a few changes: Welcome - Glenys Pillock.

Some of these component factories were brutal places. One introduced an electronic system for accurate clocking in and out, designed to replace the old card and analogue clock system where workers knew how to fudge it to punch in and out at barely-plausible times and claim maximum hours. Knowing the workers would hate it this new machine was placed inside a welded steel box, bolted to the factory wall by management. A narrow slot allowed the worker to swipe his card in and out, with no time trickery possible.

Soon afterwards they had a freak 'accident'. A forklift driver had managed to accidentally drive one of the prongs of his truck through the narrow slit in the steel box and skewer the machine inside. Getting one fork through that tiny slot required accuracy a surgeon would be proud of. If only these guys applied their skills for good, instead of evil!

A factory making suspension components was a big old production hall, girders holding up glass roofs with the light blocked out from decades of grime. Brick walls painted over time and time again, flaking like confetti if you brushed against it, in so many '70s shades of flat blue and green. Notice boards that no-one noticed, fire doors wedged open, smutty calendars and accidents waiting to happen everywhere. I loved these places.

A story I heard here was about two welders who worked side by side, often on their hands and knees, safety masks on, welding parts. One of these guys, following an earlier accident, had a prosthetic leg. He wore a steel toe-capped boot on both feet. One day, after welding in a kneeling position for a prolonged time, he attempted to stand with incredibly painful results in his stump. His workmate had spot welded the cap of the boot on his false leg to the checkerplate metal floor.

One automotive component factory was German owned. It was built in Wales when the Welsh government would chuck money at anyone who offered to build a factory and create jobs. This was a new factory, built in a far-flung corner of Europe's wettest principality, staffed by 'ard-faced ex-miners and managed by 'by-the book' Germans.

The company had no trouble filling the shop floor with local workers, but had a harder time getting experienced German managers with families to uproot. So they sent newly-qualified young German managers, graduates and other young bucks keen to prove themselves. It really was a recipe for disaster.

They were a keen user of my QA products, thanks to German insistence on inspecting everything. I missed the initial incident but an over-officious young Bavarian upset some obstinate worker and mutual hatred erupted. The workers all went to the pub and conceived a brilliant plan of revenge. The following morning management were shocked to discover they couldn't understand their workforce. Questions were seemingly answered in gibberish. They were answered in Welsh. Frantic calls took place with HQ in Germany.

Management couldn't force people to speak English, they didn't want to upset the local government or cause a PR disaster either. So, every night the young German managers were obliged to attend Welsh night school. After a few months some of them were getting reasonably competent. After six months the workers found that all management could actually speak very good Welsh. They reconvened in the pub.

The next day they all reverted to speaking English. I imagine these young managers eventually returned to Bavaria and are now, perhaps, are high up the ranks of Mercedes-Benz with the most useless qualification on their otherwise impressive CVs: “fluent in Welsh, written and spoken, isn’t it?”

Lotus, Hethel, UK

Lotus seemed to have a bad reputation for quality in the Nineties. Someone once explained that, at that time, they were sort of the masters of mechanical engineering and could squeeze everything possible out of a mechanical component. But when the digital era came, with CAD designed parts and complex electronics, they couldn't quite keep up with market expectations.

Like other fibreglass bodied cars back then they sometimes suffered with iffy electrics due to earthing problems, and a reliance on electrical components from a British supplier nicknamed "the Prince of darkness" by a Lotus owning friend of mine. They'd had a quality dalliance with Toyota but I never saw any evidence of results.

At that time they were wringing the last few years out of the beautiful Esprit, a car built by men in lab coats with fags on, on benches strewn with hand tools. I loved visiting Hethel. I reckon these guys each had enough skills (and tools) to build a complete car themselves. They didn't want digital measurement and process control. They didn't buy anything from me. I had two interesting experiences with Lotus, but not at Hethel.

Kettingham Hall was at one time owned by one of the myriad of iffy offshore companies Lotus founder Colin Chapman operated. It was a stately home close to the Hethel factory and I had an enquiry to look at a measurement application there. No satnav back then, remember. The directions I was given were quite clear and I drove my car down a muddy, narrow lane, around the back of the hall somewhere. The destination was a stable block. I nearly didn't get out of the car.

Inside the stables a small team were building race-spec Esprits for endurance racing. Their attention to detail was incredible, like nothing I'd seen anywhere else, mostly done by hand and eye. They were accurately working to tolerances way beyond what I had experienced at Porsche, BMW or Mercedes in Germany. An Esprit Sport 300 built there - an outdated car even then - held the lap record at Bathurst for a decade. After my meeting I parked back at Hethel to watch production Esprits being given a shake-down on the test track there before delivery to customers. The driver was utterly flat out. Window down. Fag on. Brilliant.

At an aluminium specialist in Worcester I looked at some interesting R&D they were doing in partnership with Lotus cars. They had a tiny, bonded aluminium tub and were experimenting with tri-lobe fasteners that were designed to deform during assembly and act like rivets. This was really interesting stuff. This was the birth of the Elise.

I later visited to give some product training and saw one of the very first test mules parked in the yard. I remember looking at the plastic panels drawn tightly over the aluminium frame, it was so pretty, so compact. So compact that a reversing HGV didn't see it. All that hard work – crunched.

Peugeot, Coventry, UK

This plant was, according to someone in their management, the most cost-effective plant that Peugeot had anywhere in Europe. You don't have to be an expert financial analyst to work out that this actually meant they made a lot of cars but spent nothing on infrastructure. Rootes built the Ryton-on-Dunsmore factory near Coventry during the war, it was then sold to Chrysler and finally to PSA (Peugeot). I could see no trace of investment of any these companies when they were making the 306 in the mid Nineties. It was threadbare.

Nearly all car modern car factories transport the cars down the production line in a kind of ceiling hung cradle. This enables the body to be articulated for ease of attaching components. Cars are presented to the component, saving workers from wasting seconds by stretching, and ensuring perfect alignment of assemblies. At Ryton the cars were attached to a cable drawn conveyor system on the floor and unceremoniously dragged along the ground from workstation to workstation. Management was often French, surly and aloof. They didn't want to come down to the factory floor to troubleshoot, and production line workers were unionised blokes from Coventry in trouble that needed shooting. The trouble needed shooting. Not the workers. You get the idea.

The French imported QA equipment from France and nobody on the shop floor wanted to use it because the instructions were in French. It wasn't like those instruction manuals you get for a new telly, printed in everything from Magyar to Mongolian but with English in there somewhere. There were no English instructions at all. Even the readouts and displays were *En français.*

We did a deal with their Management (which I never quite understood) which meant we ended up with the beautiful, if sullen, daughter of a senior Peugeot Manager from France doing some "work experience" with us. Sometimes she would be at the Ryton plant, other times she would sit at a desk in our office looking bored. She was a well-educated, sophisticated but quite dull young lady. She was also a snob who clearly didn't want to be in our little office listening to us lot trying to set fire to each other for a laugh.

I felt a bit sorry for her. One day she came to work in her left hand drive, French registered Peugeot 205 Roland Garros (the posh one). It was green. The fact that I can clearly remember the car but not her name tells you something. She told us, in her aloof manner, of her family holiday home in Antibes and that she missed the sea. I offered to take her to the beach and instantly regretted it because she agreed and I was supposed to be doing some real work. She came in the next day, as planned, and brought a picnic basket and a bag of bathing stuff, parasol, beach ball, the lot. Perfect for a day on the *Côte d'Azur*. I took her to Skegness.

For readers who don't know where Skegness is, it's an impoverished seaside town on the East Coast of England, built to entertain Victorian-era coal miners, and has been in steady decay ever since the kiss-me-quick crowd discovered cheap package holidays in the sun abroad. I had relatives who lived there once. Skegness is so ugly the seagulls fly over it upside down so they don't have to look at it. We drove up the main street and *Mademoiselle* looked out of the window and said "No! Take me 'ome! Now!" She soon returned to France. I don't think the Peugeot people ever were comfortable in England.

In later years Ryton made the Peugeot 206, a particularly shonky model that was once voted second worst car in Britain, and then Peugeot announced they were going to close the plant. Workers went on strike, asking their French comrades to join them in solidarity. They didn't.

The Ryton plant was eventually pulled down in 2007.

Porsche, Stuttgart, Germany

Confession; I recently owned a Porsche. Not a flash 911 but an old 968 Sport which, I now realise, was being made at Porsche's Zuffenhausen factory near Stuttgart at the time I was trying to sell them quality control equipment. They didn't buy a thing from me and, it's widely considered, cars made then are some of the most reliable they've ever made. Perhaps my lack of success there is linked to that. My 968 was fault-free despite my ham-footed driving. Anyway, there's little point stating the bleedin' obvious that University educated, time-served Porsche engineers knew how to test a car better than a young Englishman with sub-GCSE German and some slightly exotic test equipment ever could.

It was a fantastic place to visit. There was (and still is, albeit at a different location) a brilliant little museum where, in return for a small donation to a local charity, you could peruse Porsche models and engines, old and new, and some kindly semi-retired employee would answer any questions patiently.

A visit trackside was always possible and every worker was precious about the car on his part of the production line. The job sheet even stated the name of the person who had ordered the car. I got too close to one car in-build and was shouted at – *"don't touch that car, it belongs to…"* the guy fumbled for the build sheet on the windscreen *"…"Mr Villiams of Visconsin!"* Every single new car was road-tested on the Stuttgart ring road and the sound of the last of the air-cooled 911s rasping by at full tilt is something I won't forget. Nor will I forget the last of my futile visits trying to show them how to do things differently.

I was wearing a new black suit. One of those slim fitting, crease-free travel suits. Quite thin, but good quality. I was explaining something to a patient barrel-chested Swabian QA Engineer when I glanced down and spotted a bright blond hair in my lap.

We carried on talking and I tried to nonchalantly brush it off. It never moved. I looked down, and so did my customer. There was no hiding it now, curly blonde hair on black suit. He saw it, I saw it. I was trying to convey the image of a meticulous measurement expert. I took aim and flicked the hair, without pausing the conversation. Again it didn't move. I then yelped as I plucked the pubic hair that was sticking through my thin suit. I don't remember how I disposed of it, but I do remember he refused to shake my hand when I left.

Pictures

With thanks to the British Motor Industry Heritage Trust (who own the copyright) we have a selection of pictures. Below we have a Noel Edmonds cutting a cake to celebrate the end of production of the classic Range Rover at Solihull in 1996. Notice the sullen faces in the background and how the icing on the cake isn't straight.

I asked him for a quote for this book. His Manager said "Unfortunately Noel is too busy at the moment". Quentin Willson wasn't too busy, but wanted £1500.

Rovers K series engine was actually rather innovative. The block was made by pumping liquid aluminium into a chemically bonded sand mould from below. This process gave advantages in limiting the occurrence of unwanted oxide inclusions and giving a far better yield than traditional casting methods. Clever stuff.

The man above is working in the low pressure sandcasting department at Longbridge. The protective headgear ensures he may listen to Slade on his Walkman uninterrupted.

When BMW bought Rover there were a number of bonding days. Pictured here, in 1994, is the then BMW Boss Bernd Pischetsrieder (left) and MG Rovers John Towers (right) looking uncomfortable in the mud at the Gaydon test track. Note Pischetsrieders badge falling off.

If you put this page close to your ear and listen carefully, you might be able to hear that Defender rusting. Pischetsrieder is a distant relative of Mini designer Sir Alec Issigonis, which might go some way to explain his incurable Anglophilia.

This is a photograph of an Austin Montego. The chap kneeling is showing a group of visiting school children how to make prototype panels from clay on a 'design in action' day. Or maybe he's just clagging up a wing.

Pictured below: Rover 800 windscreen installation robots. Yes, I know these pictures are on their side. Or maybe the factory was.

I love this photograph. It epitomises Rover at Cowley for me. It shows two men with slightly bewildered expressions using a handling machine to help fix an exhaust to a Rover 75. I imagine the conversation as follows; "*ow's it work?*" says one, "*I dunno*" says the other "*let's see if we can knacker it when no-one is looking*", before one of them trips over that cable on the floor.

Roughly 80,000 Rover 75s were made at Cowley, before it became a MINI factory under BMW, and 75 production was moved to Longbridge where they made another 151,000 or so. Cowley built cars usually have black painted sills. Despite everything I'd still rather like to own a Rover 75. Don't tell anyone.

Your car, here

Not all cars are Friday cars. You might well own an example of a car made in one of the factories mentioned so far and be very happy with it. When I first started to write about 'confessions from quality control' I was bombarded with angry two-finger typed emails from people delited [sic] with their Carcompany Whatever GLi Deluxe. I'm very happy for them.

As mentioned in the introduction I have no axe to grind here and would be happy to own just about any of the cars mentioned. Apart from an MG RV8, perhaps. What I describe in this book are probably just isolated incidents. But just because your car hasn't been mentioned doesn't mean that the factory is infallible. It just means that perhaps I never visited, or if I did then perhaps I don't remember anything amusing enough to give it it's own chapter in this little book … or the lawyer says I'll be shot if I print what I'd like to say.

I never went to TVR, sadly, and if I had I would expect there'd be a chapter or two on their Blackpool factory based on my experiences of driving them. There's very little Renault in this book either although interestingly I have no fingerprints on one finger after extinguishing an electrical fire in the family Scenic with my kids still inside it; it's a brand I hate, but I never visited the factory.

I do have a few other little memories of other car plants which I'll share here in case Angry of Aberdeen thinks it's unfair I haven't mentioned a lemon he once owned.

Opel have a big facility at Eisenach. I was hoping to leverage on the fact that I had supplied the Vectra and Frontera factories in the UK with equipment at the site in Germany where they made Astras. At the security desk was a wanted poster with an artist's impression of a man wanted in connection with a series of sexual offences in a nearby park. He looked a bit like me. This didn't help. It was neither a happy or productive time I had there, although they seemed no less slapdash than their British counterparts.

My abiding memory of Audi, at their Neckarsulm plant, is of an incredibly slow production line. So slow was the A8 line that I thought it wasn't moving at all. Each worker's station on the line was spotless, some having framed pictures of family, or a houseplant of some sort. They seemed relaxed and a world away from the high-tempo environment of the Japanese managed plants. Things were checked to national and international standards (Germans love to standardise things), documentation was immaculate and their test and measurement instrumentation was the very best. A boring place for book research. They're boring cars, too, aren't they?

Seat, part of the VW family in Spain, bought very expensive equipment and simply didn't use it. It's never nice to generalise but I too often found that the quality results were a reflection of the perceived national characteristics of the workforce. Standards can be written, instructions given, machines programmed and measurements made but it always comes down to the attitude of the workforce. Where factories were built from scratch with a new workforce, such as the Japanese ones in the UK, things were very well controlled. Others that had seen many owners over decades of production, such as Peugeot's Ryton, were something else.

I found Russian Ford workers wanted to know absolutely everything, eager to learn, but were hampered by complicated local politics, Brazilian Ford workers intelligent but lackadaisical and too many French car workers I met seemed to have a 'not my fault' attitude.

It always seemed to rain when I visited Aston Martin's old factory in Newport Pagnell. I once signed the visitors book underneath the name of Prince Charles, which was impressive. I saw unfinished bodies being pushed about on trolleys outside between buildings, in quite rotten weather, which was less impressive. I can't really comment on their quality at the time with any accuracy as I didn't visit often, but an outspoken acquaintance once worked there and once described it thus; "*An awful place to work - TWR smartarses, jaded 'could't give a f**k' Jag Coventry drones and six fingered local bumpkins. Plus some pretty unpleasant Ford people.*" He later left under a cloud. A metaphorical one, not an actual one.

I'd quite like to write a chapter or two on a German motor manufacturer. Not that there were many instances of shop floor shenanigans, but because of management corruption. I knew something was odd when I found some buyers would give you their business card and it had their home address on the reverse side. Who would give a salesman their home address? People expecting gifts, that's who. I learned the phrase '*großen hand unter dem tisch*' which roughly translates to 'a backhander' in English, and '*schmiergeld*' meaning 'lubrication money'.

One particularly bent buyer said we had to give him either a big discount on our products so he looks good to his boss, or a big 'personal discount' to make it personally worth his while. I found this really hard to stomach and as a result of that, and a sometimes underhand competitor, we didn't do much business with that particular German motor manufacturer. Maybe if I sell enough copies of this book I'll use the profits to cover the inevitable legal fees needed to cover my arse and write it up some day.

I'm sure things have changed now but it made for very interesting business back then, and I don't doubt that *all* manufacturers have their 'moments' in quality control. So Angry of Aberdeen, let's get back to what this book is really about: the calamities. And I caused a few of my own…

Mitsubishi, Bristol, UK

Mitsubishi don't have a factory in the UK, but when I glanced down off the M5 near Avonmouth I saw a huge amount of new, unregistered cars, parked up. I decided to take a look. Security was tight. It was the port where all new Mitsubishi cars were unloaded from ships from the Far East, where they were made. Still, always chasing a sale, I found an office on site and asked what kind of inspection they did on these cars at time of unloading.

The manager was short of time and a bit annoyed I'd arrived on site trying to flog him things when he had a ship to unload. "*The cars are perfect*", he told me, "*they're made in Japan. Now f*ck off*". On f*cking off I managed to reverse into one of these perfect cars. In my head, whilst driving off, I did a '*it's just a scratch, probably nothing*' bit of thinking, too scared to go back and tell the narky bloke I'd just dinged one of his cars.

As I drove off I looked in my rear view mirror, wondering if anyone had seen me. As I did that I managed to take the wing mirror off of another perfect car, and when getting out to inspect the damage, the wind caught my door and smacked another.

I hereby, after twenty years of guilt, apologise to Mitsubishi motors for damaging three of their brand new cars and running away. I also apologise to whoever took delivery of their 'perfect, made in Japan' car and wondered why it looked like it had been in a car crash.

Fiat, Turin, Italy

I had been warned that Italian business, particularly in the car making world, is all about relationships first and products second. I heard that Fiat was crazy about automation, new technology, and were prepared to take technical risks that some of the other car manufacturers would not. They, like others, were so unionised that they had more workers than they needed. Any machine that would save time was welcomed, even if they didn't trim their workforce once they'd implemented this time-saving equipment.

Some of the best controlled factories I had dealt with still relied on a man to double check what the machine had done. Men can spot things machines cannot, can rectify it on the spot without reprogramming, and while this adds a layer of cost it also adds a layer of security. Fiat seemed to have not just one man checking, but another man checking him, and another checking him, and in the end there were so many layers of checking that most of them didn't bother because someone else in the chain would spot any defects. Probably.

Just like Fiat, our business in Italy relied on a chain of largely unreliable people: Distributors, agents and middle men. The end result was that we got paid four to five months after delivery and so many people had taken a cut from the deal that we made no profit. As long as everyone was friends and had a plausible story, business was good, and finding someone who gave a stuff about how things were really made was nigh-on *impossibile*.

I recall a long meeting with a wiring loom manufacturer, taking a factory tour and spotting an obviously defective part, frayed ends and broken plastic, complete with a QA approved sticker on it, ready for delivery. When I pointed it out I was shrugged at and told "*ees only for Fee-at!*" Someone else would have to fix it. Fiat was frustrating and Turin was crazy.

I once had my bags stolen from the boot of a locked car outside the Fiat factory. Someone suggested the thief had run in a certain direction and we jumped in our jemmied Lancia Thema and gave chase. We found a carpark, ankle deep in fly-tipped litter, and people living in transit vans on bricks with naked children and barefooted women with their hands out, begging from us businessmen. They were welcome to whatever they'd stolen, it was a horrible existence and as a father myself it really hurt to see the children.

My travel insurer said I needed a document from the Police so I went and queued up at a Police station, struggled to fill in a form for a bored Policeman with a uniform an African Dictator would consider OTT. He never smiled, barely answered my questions in my nightschool-learned Italian and never took his shades off. I spent ages filling in this form which just needed his signature.

I passed him the form through a hatch and was told to wait a minute. I saw him walk just out of sight and saw, in a reflection, that he was stood smoking. Then he picked up a newspaper and read for 15 minutes or so. I knocked on the desk and he ignored me. Eventually I snapped and stuck my head through the hatch and shouted "*sign my form you lazy b*stard!*". After being forced to apologise, and wait two more hours, I got my signature. I wasn't very good at this Italian relationship thing.

Toyota, somewhere unpronounceable, Japan

Following a good relationship with Toyota in Europe we managed to send some of our latest equipment to Toyota, somewhere unpronounceable, in Japan. It was, looking back, a ridiculous thing to do.

Somebody needs to explain how it works, train the user, prove the results are correct and do years of bowing and nodding before Toyota would even consider buying anything. Also, the equipment used a new type of electro-mechanical hydraulic system to work and was a bit 'frizty'. We couldn't afford to send someone from our factory along and didn't have anyone in Japan to help, so it was just crated up and shipped off.

They were given my mobile phone number so, considering the time difference, they could call and get technical help at any time. I heard nothing for a month or two and then, at some late hour, a strange looking number called. "*Frames!*" said the voice. It was Toyota Japan, and they were having problems. I could make no sense.

The voice was attempting English and it sounded a bit urgent and desperate. "*FRAMES!*" I asked what they were doing with the equipment and understood they had put it outside the factory. What was it doing there?! "*ROTS OF FRAMES – IT ON FIRE!*".

We didn't ask them to return the equipment, and didn't attempt to sell anything to Japan again.

Daewoo, Worthing, UK

It wasn't a bad idea. Today so many cars are bought on finance, or leased, and treated as white goods. In some respects Korea's Daewoo were ahead of the game. They offered very cheap cars sold at a fixed price via no-hassle dealerships. They could afford to do this by buying old designs from GM, giving them a facelift, rebadging them and building them on the cheap in Korea. This year's Daewoo Nexia was yesteryear's Vauxhall Astra.

Daewoo understood that the facelift bit was critical. Buyers wouldn't knowingly want to eat yesterday's leftovers, so Daewoo peeled the best-before labels off and microwaved it for an extra hour or so, to be on the safe side, before serving. The facelifty bits were done at a huge new technical centre in Worthing where Daewoo had recruited a new team of engineers to help their dog's dinner seem more palatable.

They had quite a budget to get things off the ground. They had taken delivery, in great secrecy, of the first examples of Daewoo models for a technical review. To see if they had been built to the standard expected by European buyers. I was invited, with my case of inspection instruments, to demonstrate what to look for and how to test.

On arrival I was greeted by an engineer who proudly told me he had been someone senior at Ford and had been poached. At a meeting I met lots of 'senior engineers' who all told me they knew my (rather special) equipment, and had used it before, and were all ex-Jaguar, ex-BMW, ex-Rolls Royce, etc.

We were a small company. I knew who my customers were for what products and knew that none of these guys could have seen this particular equipment before. Why were they trying to impress this slightly shambolic twenty-something salesman? I concluded that these "senior engineers" were nothing of the sort. They were far less qualified, if at all, and Daewoo (who had no history of car building) had probably employed them on the strength of over-inflated CVs at a time when good engineers were in short supply.

We started with some tests on an Espero. This is a model that was designed in the mid 1970s by GM and best known in the UK as the Vauxhall Cavalier MK2 (or an Opel Ascona C to continental readers). A twenty-year-old design being sold as a new car. Firstly, there were seatbelt fixings loose. Then door hinges. Not straight. Some bits were rusty already. Each item we tested was nowhere near the factory spec. None of these 'senior engineers' had a clue what to do about it.

Daewoo slowly unravelled. Their fixed price models weren't quite fixed price if you had a trade-in, as different dealerships valued part-exchange cars differently. The motoring press quickly got wind that these cars were old Cavaliers with Listeria. In 1999 Daewoo were bankrupt, owing a spectacular $84 billion. Wanted posters were put up for boss Kim Woo Choong who had Interpol after him for embezzlement and fraud. He was eventually caught and sentenced to ten years in jail.

In 2001 the Daewoo Worthing site was closed, and there were c.16,000 of their Esperos on UK roads. Today there are only c.70. White goods indeed.

VW, Hannover, Germany

Pedro and I boarded the ferry to Bremen with a van packed full of our latest equipment. An elderly German sales agent, let's call him Günther, had managed to get us an appointment at VW's van factory in Hannover. VW were a big target for us and we jumped at the opportunity to do a product demo to their QA team. 24 hours on a ferry with a hard-drinking Geordie warped my mind a little, and I became convinced that VW would welcome me with open arms, like a modern-day Major Ivan Hirst, the chap credited for rebuilding the VW factory after the war.

We disembarked, remembered that foreigners really *do* insist on driving on the wrong side of the road, and collected Günther. Three of us crammed into a small van, heavy with equipment. I hadn't worked with Günther much before. I had heard that, after the war, it was easy to get a job as a salesman in Germany. Everything was being rebuilt and everyone had money to spend. Günther was of that era and, looking back, we'd had way too much faith in his ability to act on our behalf at VW.

It's no secret that German companies like this were rigorous in following standards. We set up to make a number of measurements to a well-known standard, knowing that we'd get slightly different readings. This would usually spark a detailed discussion of how and why and we could point out how our equipment was superior to their current inspection set up. This required careful words, excellent measurements and no salesman's blather.

The lab was immaculate: we were offered coffee and set up our system. Günther insisted on doing the talking, even though I asked him not to. He was a bit deaf, which didn't help. He had no idea of what he was looking at. Every measurement we made was followed by "*Ja, gut*" or "*Ja, in ordnung*", even when it wasn't. Measurements became more erratic but instead of pausing and letting the customer understand what we were doing, he just droned on "*Ja, in ordnung*". People were shaking their heads.

It was a disaster. I wanted to kill Günter. We'd travelled so far and had no chance to explain or discuss anything. Just "*Ja, gut*". The VW people wandered off, disinterested. I could see what coming next. Pedro squared up to punch Günter, I could do nothing and just as Günter realised what was coming he finally stopped talking, his jaw dropped open and his false teeth fell out, and broke.

We packed up and drove back to port, wedged in the van, in silence. Günter fell asleep and I had to stop Pedro from stealing his hearing aid. We dropped Günter off. "*Zhang you, zhat vosh a great zhugzhess*!" he said, toothlessly, brainlessly. Even Major Hirst couldn't have helped him.

Ford, Dagenham, UK

Some readers might be old enough to remember when the Ford Cortina was referred to as the Dagenham Dustbin. I always thought this a little unfair. I never owned one, but my Dad had a few when I was growing up and they were alright. Like many other children of the Seventies I had many happy holidays in the family estate, burning my arse on the black vinyl seats.

I'll confess that I didn't visit the big Ford factory too often back when I was a sales rep because the place seemed such a dump and the man in Quality scared me. We shall call him Mr Tefal, because he always reminded me of a small, round kettle that was only just able to keep it's boiling rage bottled up. He was constantly on the verge of a heart attack. Fists clenched. Everything he said was through gritted teeth, if you could get hold of him at all. I once spotted him eating lunch. He was having chocolate and vodka. He was scary.

It was the very late Nineties and, sadly, the writing was on the wall for car production at Dagenham. This place once employed 40,000 people but were down to assembling just one model. They made the Fiesta and the Mazda 121 (identical apart from badges and a pointless bit of trim). It was an unhappy time to try and sell any QA kit because equipment budgets were generally only allocated with new models and the Fiesta had been reluctantly slung down the line there for some years already. Rumours (later substantiated) were that Dagenham wouldn't be making cars in the future and they invested accordingly.

The only piece of quality control kit from our company that Dagenham owned was a decade old and a bit knackered. I was surprised to get a phonecall from Tefal. "*Is your fackin' machine year two fahsan' compliant?*" A very strange question. I went to see him at the massive sprawling site, off the A13 east of London. I parked miles away as my company Honda wasn't permitted to park in the visitors bay.

I don't know if it was a co-incidence or if there was some automotive apartheid going on, but I used to take a shortcut between two assembly areas and noticed something odd. One area was manned by Black and Asian workers, the other by White workers. The atmosphere was tense and the place was run down. It transpired that Tefal had been given a budget to upgrade any of his ancient QA inspection equipment that wasn't year 2000 compliant. I am probably going to hell for this, but I told him that the ancient box he was using wasn't. It didn't even have a clock but he didn't know that and was too busy randomly shouting at line workers to notice anyway.

I told Tefal that the old product's replacement *was* Y2K compliant and we had one we could deliver before all the clocks went pop at the end of the millennium. It was complex, hugely expensive and in the wrong hands you'd get more accurate readings from a gypsy palm-reader. He soon placed an order but only on the agreement that I would come back and train his QA team how to use it properly because, and I remember this quite clearly, "*I fackin' 'ate 'avin' to tuwk to those cants*".

I'm many words into this chapter and still haven't touched on Ford's quality. This is never going to be a comprehensive site assessment, backed up with graphs and stuff, it's just my own rambling observations – but I could tell a lot from how people used my equipment. Those that knew how to measure, who understood the readings and were intelligent enough to implement appropriate changes in production were great. They made good cars. That kind of person I never met at Dagenham.

I arrived with this very expensive equipment and was given a grubby corner in the factory to train their quality inspection "team". One man turned up. A pot bellied, oily little man who wore overalls wide open to the navel, showing a magnificent chestwig and the occasional hint of nip. He had strange hair. He was the man who helped to check that Fiestas didn't fall apart, or rust, or explode. Tefal turned up to "*keep an eye on the little cant*" who I shall call Bob, as he reminds me of Bob Fossil off *The Mighty Boosh.*

The equipment would make a series of measurements from which you could certify how well the car was being made. I don't want to go into details, as explained earlier in this book, as it's a bit niche and the company is still in business. Their usual equipment was robust, accurate and user-friendly. This thing, the most expensive thing I could sell, was not. It relied on a lot of programming to work correctly and one mistake meant your results were cack. I could not stress this enough.

I started to go through a long setup list and made sure Bob Fossil and Tefal understood it. I was five minutes into the training when a visibly impatient Tefal exploded "*jast press fackin' start!*" Every vital instruction was shouted over with "*jast fackin' start the fing!*" I was already out of my comfort zone when Bob Fossil nonchalantly pulled out a mobile phone from inside his overalls and called directory enquiries. I was stunned.

Tefal was shouting. Bob was slowly asking for the number of Boots in Barking. As he was connected he calmly looked Tefal in the eye and told him he was on his break. Tefal, now raging "*snot fackin' break yet!*", grabbed £2000's worth of test instrument and blindly threw it, still ranting, missing a chap who was passing with a tea trolley but hitting his urn with a clang. Bob was talking to Boots and stroking his fringe. "*No*", he drawled "*it's not auburn, it's deeper, more of a burnt ochre. It's usually in a bigger bottle. That's the one. £6.75? I'll collect later. No, madam, thank YOU.* " He was ordering hair dye in company time. Tefal screamed one last, lung-bursting "*YOO CANTS!*" and stomped off. Bob never even flinched. He said it was his break time.

This seemed normal for Dagenham. I had the feeling then that my QA equipment, still in it's protective packaging, would never be used. Perhaps it was thrown in the mucky Thames that runs behind the site, or car-booted by a pilfering worker.

I apologise that this chapter (and, indeed, probably half of this book) reads like a work of total fiction. Like a sweary episode of *On The Buses*, or *Carry On Car Factory*, but twenty years ago this is what I encountered at Dagenham. The Mk4 Ford Fiesta, it now seems to me, was a bit of a rotbox. But the guy who inspected them probably still has the shiniest, most lustrous ginger comb-over imaginable.

The perception of quality and Japanese production terminology explained

The perception of quality is interesting. To some it means the look and feel of the touch points and materials used in a car. This is why the VW group, for example, took huge leaps forward in the Nineties when they gave us nice plastics, damped grab handles and dashboards that lazy motoring journalists said were 'hewn from granite'. But that's not to say they were assembled properly, or were electrically or mechanically dependable, and to many people this is their measure of quality – is it reliable?

Japanese cars have a reputation for dependability but perhaps don't always score highly on the tangible bits such as the feel of their plastics. Toyota have used precisely the same type of electric window switch on many models, millions of times over, because they know it works perfectly. They won't change it for the sake of cosmetics.

Japanese workers will perform the same task endlessly for years without getting bored, making their production process extremely repeatable and therefore resulting in a very consistent and dependable product. Many Western carmakers take their lead from Japan (whether they admit it or not) and as a result we have a little lexicon of Japanese phrases creeping into use, some of which you may have heard. I shall explain them as follows:

Kaizen – this means, literally, improvement. In a production environment that might mean elimination of waste, better efficiencies, or (Rover) getting the canteen to serve their soup with less lumps in it.

Kanban – meaning signboard, this is used in describe lean manufacturing and, in particular, 'just in time' delivery of components which are monitored using Kanban cards helping trigger supply to the line.

Kaikaku – means radical change. Kaizen (above) is a gentle, constant system, Kaikaku is something with a bigger and more immediate impact. It is graded from perhaps replacing a worker with a robot, through to the "Radically innovative - Operation close" stage.

Hansei-kai – is a meeting where, even if a project went totally swimmingly, you reflect and self-analyise to the point where you eventually find a fault and therefore potential to make an improvement next time. A meeting for masochists, basically.

Mottainai – is a phrase used to express regret at producing avoidable waste [insert your own Austin Allegro punchline here].

Poka-Yoke – Meaning, roughly, 'avoid mistake' by means of making a component only fit one way. Like those little plastic tabs on cable connectors that invariably snap off.

Ishikawa diagram – Named after its creator, Kaoru Ishikawa, the diagram (sometimes called a fishbone diagram after its shape) shows the causes of a specific event. Ishikawa was chairman of the editorial board of the monthly publication Statistical Quality Control, which sounds like a laugh-a-minute read to me.

5S – Those five are *seiri, seiton, seiso, seiketsu*, and *shitsuke*, which translate to sort, straighten, shine, standardise and sustain. 5S is a method of organisation that is completely unworkable for those with a lisp.

Vauxhall, Luton, UK

IBC vehicles are an offshoot of GM. In Luton, near the airport, sits the huge IBC factory that spewed out Vauxhall and Opel Fronteras in the Nineties, where it previously (and subsequently) made vans. I knew that if I could get a meeting with a particular senior production engineer (we'll call him Dave) and demonstrate my kit, I'd most likely win an order. I accept that this chapter isn't a comprehensive assessment of the quality of their vehicles, but I'd like to think it's a fairly accurate insight based on what I saw.

The day started well, I was early. Too early. In fact I was so early I had time to drink three coffees in a cafe near the factory. I then needed a wee. I didn't want to arrive at my meeting and ask for the toilet so I drove around looking for a public convenience. I found one. Flies to my cheap suit undone I let loose, nothing could have stopped the flow, and at that point I noticed two obese Asian men having sex in the corner, staring at me. The day got worse.

I had been on site before but in order to walk about the production line without a chaperone you needed a safety briefing. I sat in my piss-soaked trousers next to a sullen Eastern European chap in high-viz. We were set a little test to ensure we had been listening. Igor leant over and asked me "*wot does mean word we-hickle?*" He was a welding inspector. We both got our passes and I walked down the production line to have my meeting with Dave from QA.

Back then we didn't have camera phones and cameras were not allowed on site. To this day, I regret I didn't get a snap of one thing I spotted there that summed the place up so well; at the end of the production line was a vandalised sign where some wag had added a '**C**' on a sign, which then read: "Warning! Vehicles emerging from **C**left"

The place was dirty and disorganized. A huge expanse of car park looked like the despatch yard but was full of cars that had come off the line and needed rectification before being shipped off to the dealers. Most had a bar-coded label with printed description of the fault to be rectified. Others had bits of tape stuck to the windscreen with marker pen messages such as '*brakes f*cked*' and one which stuck in my mind, and consider this is supposed to be a document to instruct someone how to rectify a fault, said simply '*is sh*t'*.

I had been told where to find Dave, and as I walked alongside the production line I wondered what kind of person could manage such mayhem. The Frontera had a poor reputation for quality and surely I would make a sale when he saw how my equipment would improve their productivity. It had been rushed into production as GM wanted an SUV in their lineup; IBC had got the short straw.

As the line ground to a halt (an accident, industrial action, who knows?) I got closer to the section where Dave worked. I asked a line-worker where he was and was pointed towards the body of a half-finished five door Frontera. Slumped in the drivers seat, boots on the dashboard, smoking a fragile looking roll-up and simultaneously eating a limp sandwich, sat Dave. "*Fag break*" he said, dropping crumbs and ash everywhere. "*Come back later*".

The Frontera came fourth from bottom in a JD Power satisfaction survey in 2001. I'm surprised it did as well as that. In defence of the Frontera – and I'm quite sure many people love theirs - one did set a record in 1997 for the fastest circumnavigation of the globe by a car in 21 days.

I think the final words of this chapter should go to a frustrated member of the owners club who said the Frontera was "*built by people who weren`t really sure which end of a spanner to use*".

Kit Cars, and Triumph, Hinckley, UK

My company car came with a flip-front Motorola car phone and, as I was now addicted to this job, I was glued to this phone chasing new customers and projects. There were no laws against using the phone while driving either, and far fewer speed cameras. I drove huge distances every day, flat out, chasing everything. The car factories and component factories were good customers and I started to chase even-more obscure targets.

The UK had quite a sizeable kit-car market in the Nineties. It was partially a throwback to the Sixties when people wanted to save tax on new cars and would build them themselves. The Government closed this loophole by stating, somehow, that component cars should not come with an assembly manual. Lotus, ever inventive, continued regardless and once delivered cars with a 'disassembly' manual, encouraging people to read it backwards.

In my time, kit cars were (and still are) mostly either copies of Lotus's iconic 7, or copies of otherwise unaffordable classics, such as the AC Cobra. These companies would sell kits, or factory-build cars; spotting the words 'car' and 'factory' in the same sentence would invariably be enough to have me charging up a motorway to knock on their door. I'm not going to name any - some are still in business - but nearly all those I visited were unlike any regular car factory.

You only get one chance at making a first impression; I recall visiting one Lotus 7 lookalike 'car factory' to find the front door was on upside down and opened outwards. It even had an upside down letterbox. Everything felt like a bodge. To this day I have an inherent distrust of kit-cars because of that front door.

Triumph, on the other hand, is a real success story. Rescued by housebuilder, John Bloor, they've built a large and thriving business building motorbikes. It was so organised, when I visited, so efficient, that it didn't feel British somehow. They developed and built their own engines, many component parts coming in from overseas. The assembly line was quick-moving and made c.750 bikes a week. Everything was tested, bikes were thoroughly checked pre-delivery and it was all very impressive.

No 'confessions from quality control' here; they knew what they were doing. But in March 2002 they had a colossal fire. At one stage 100 Firefighters were in attendance, and half the place was ruined. Triumph recovered and rebuilt, thankfully, but I harbour this strange hope that there *was* a bit of old-school Britishness lurking in there somewhere after all - a stray fag, or someone hotwiring the canteen sandwich toaster, for example.

Ford, Southampton, UK

Sadly, it's gone now, like so many other factories the Ford plant near Southampton has since closed down. They were, in the Nineties, a regular customer. I used to offer regular on-site training for customers like this to help ensure they got the most accurate results.

The man at Ford Southampton was never, ever available. Sometimes I would catch him on the phone and he'd agree to see me, and I'd trek all the way down to Southampton to find he'd been called off site for 'an urgent meeting'. Now, as someone who isn't averse to swinging the lead himself, I had the feeling that perhaps 'Ollie' was up to something not related to assembling builders vans. Someone took pity on me after yet another aborted meeting and let slip that Ollie was, yet again, away racing his Lotus.

Ollie raced his Esprit hard and, as Lotus fans know, they do need a little bit of care. Ollie had bought his car new and it was still under warranty. On collecting the car from yet another repair it conked out in the New Forest due to an electrical glitch. And then burst into flames. Hampshire Fire Brigade were called and the car was toast, and so was a swathe of the New Forest, apparently.

Ollie was involved in a protracted argument over getting a replacement from Lotus and was given a courtesy car from the factory to tide him over/shut him up. Lotus didn't know Ollie was racing his own car - or that he intended to use this courtesy car as he would his own by taking it to Goodwood for a few hot laps. This courtesy car actually belonged to Mike Kimberly, boss of Lotus, who was away on holiday.

After a few laps a senior employee of Lotus spotted his boss's car being raced, dashed down to the pit lane and snatched the keys. The brakes were already kippered. Ollie was rarely at Ford because he was too busy enjoying automotive adventures like this; Lotus eventually gave him a replacement Esprit and the boss of Lotus later left through ill health. I doubt the stress of this episode helped.

Ford closed the Southampton plant in 2013 and moved production to Turkey. It also marked the end of over a century of vehicle production in the UK for Ford. Ollie was given a job at Lotus.

PSA (Peugeot/Citroen), built to a budget. France

Working with a sympathetic French tier one supplier, who worked almost exclusively for PSA (the group who owns Peugeot, Citroen and others), I got a fascinating insight into why so many of these tier one companies were in a financial mess in the Nineties. After being squeezed relentlessly on price the buyer took me to the canteen for lunch where a large glass of red wine and passive smoking were compulsory.

PSA, I was told, were building cars to a price. They were dictating to suppliers, like them, the price they would pay. Suppliers would receive an enquiry, for example, for the supply of an exhaust based on an estimated quantity over the lifetime of the vehicle. Once the part has been tested and approved (PPAPed to use the lingo) it was almost impossible for the car manufacturer to change it, and of course if you've built a factory to build exhausts for some 'orrible French hatchback then you can't suddenly stop and start making yoghurts instead, or whatever other profitable product takes your fancy. Contracts are signed, drawings approved and delivery starts to take place.

It isn't as simple as cost + profit = selling price. Well, it probably was, once, but I was told PSA (and I am sure others did the same) would insist on a cost breakdown. And that breakdown had to include everything in minute detail. Then some twenty-something female Eastern European buyer with killer cheekbones, stone-cold eyes and a point to prove would then dissect everything. Your sub-suppliers would be visited, your financial situation analyised, your quality standards rubbished and above all, your profitability checked in minute detail to ensure you're not making more money than they think you deserve.

Invariably they would never need as many exhausts as originally anticipated. For example, Renault would break even on their Avantime model by making (and selling) c.20,000 per year, but only made c.8,000 in total - ever. So the supplier's costings would be way off, economies of scale lost and that slender profit would become whopping loss. It gets worse. Komrade buyer would then visit. She would inform you that they have had to slash the cost of the car because no-one is buying it, and therefore suppliers have to slash their costs too. It was a relentless squeezing and caused pain throughout the supply chain. On announcing to the workforce that cost reductions were needed they invariably went on strike, protected by France's generous employment laws. It all got a bit *British Leyland.*

It actually went a step further, so said Pierre. Using the detailed cost analysis from the supplier, when launching a new model to replace the one that didn't sell, the car manufacturer would almost dictate the price they were willing to pay. They had a cost model of the entire car. Marketing would say they can sell the car for €12000, of which (say) drivetrain is 15% of the car, and the exhaust 3% of that, meaning they'd pay 25p for it. Or else you can use your exhaust factory to make yoghurts instead. And so car components, and therefore the entire car, would be built to a price.

In the JD Power UK customer satisfaction survey of 2001, which judges vehicles on subjects such as quality and reliability, I don't think there was one PSA group car in the top twenty. On the plus side, since my meetings with Pierre, I developed quite an appreciation of red wine.

Annoying the Bavarians

I had gotten the job of dealing with European customers on account of having done a half-decent job in the UK, and because my ability in foreign languages was only marginally less terrible than that of my colleagues. Despite my Germanic pen-name (a nod to a favourite character in the BBC black comedy, *The League of Gentlemen*) I am English.

Germany was a big potential market for us. We were a small and relatively unknown company with new technology, selling in a market with some strong domestic competition, to huge car companies. We were trying to tell BMW, Mercedes Benz, Porsche and VW that they probably weren't building their cars properly unless they used our kit. It was a massive challenge. Despite all that, they sent me.

My German wasn't that bad. I'd spent time at Siemens in Saarbrücken as a teenager where I caused chaos. We were supposed to start at 6am and I was never there before 9. My boss hated me but after I literally bumped into him in the street on a Saturday morning, leaving a sex shop with two very large carrier bags, he left me alone for fear of blackmail. Even after a fellow apprentice and I pranged his BMW 635 company car. Some years on, selling measurement equipment, I remembered the basics, and enjoyed visiting.

In Germany I used to like staying in countryside inns, as they were cheap, clean and very hospitable. There were often interesting people about too.

The UK used to have 'half day closing' where shops would close on a weekday (remember that?). Parts of Germany do the same and call it 'Rühetag'. I was staying in an inn in Bavaria and because it was 'Rühetag' the place was empty. I used to buy the local paper and read it to practice my language. I remember being so relaxed: cold beer, homemade bread, cheese and ham, sun on my back after a long day arguing the toss at BMW.

An old, local chap asked if he could sit with me. He had half an ear. He was quite friendly but I had a feeling he was going to spoil my evening. After a few words about something in the newspaper he heard my accent and asked if I was Dutch. No. It's perhaps unusual for Brits to speak German and I'd confused him. Danish? No, I'm English. At that point he went quite mental. The Bavarian dialect is hard enough to follow but the gist of his rant was that I was an English arsehole and I had shot his ear off.

I told the guy I was born roughly 30 years after the war ended, offered him a beer, and he calmed down. He was just like my own Grandad; Opinionated, ignorant but interesting and I would have loved to have shared a beer with them both together. Actually, I suspect my Grandad would have shot his other ear off for a laugh.

The company owner was a great Engineer but sometimes weak on detail. But he was great fun. A barrel-chested Englishman who usually wore a tweed jacket and shoes a homeless person would refuse to touch. He was also very, very posh. I recall a particularly unsuccessful visit to BMW that coincided with Oktoberfest. We sat together in neat rows on wooden benches, eating goulash, drinking beer, and listening to an Oompah band on stage. But, somehow, all the fun had been organised out.

The boss had a gleam in his eye and dog-eared tennis ball in his hand. “I bloody hate this music” he said. He threw the ball at the bandleader, but his shot fell short and landed in a fellow diner’s goulash, splashing the lot everywhere. We ran off. This incident sort of sums up our results of trying to do business with BMW.

BMW were entwined with a local competitor and we could hardly even get a meeting with them. One meeting was convened but held at our competitor’s offices, and we had a weird three-way conversation between us. I had to threaten to pee in a pot-plant on the receptionist’s desk before they would let me use their toilet. I look back and wish we’d thrown a few more things at Bavaria’s Oompah bands

Hummer, Mishawaka. USA

The spellchecker on this PC, as I type, wants to change Hummer to bummer. Mishawaka confuses it, too, and it confused me when I visited the Hummer factory there in the state of Indiana.

It started badly. The American sales agent, who I had never met before, collected me and had just been left by his wife. He was in tears. We drove along in some colossal SUV thing with tears streaming down his face. Everything seemed to make him cry. A song on the radio was Jane's favourite, that made him cry. We stopped for a burger, I unwittingly ordered what Jane would have had. That made him cry too.

We arrived early at Mishawaka [how do I switch this spellchecker off?] and ended up in a Yankee Candle shop to kill some time. I wandered off trying to lose my tearful colleague when I was cornered by a shop assistant, fascinated by my accent. "*Say sumin*" she said. She wouldn't leave me alone. "*Say sumin else!*". After fifteen minutes of polite replies I'd had enough. "*B*llocks to this*". She loved it. "*That's so cute!*" I collected my sidekick, sobbing over a fragrance that was Jane's favourite, and we went to the Hummer factory nearby.

The Hummer was designed as a military vehicle, with a six-point-something litre engine, a ground clearance of 40cm and the ability to ford 76cm deep water; it was the perfect vehicle for invading foreign countries on a whim. The Hummer H1 model was the brainchild of someone at General Motors who thought it would make a good road car. The Mishawaka plant was gearing up to make these things and I am sure my approach to quality control was as unappealing to them as their vehicles were to me. I don't think they used the words 'commie pinko' but the sentiment was there.

Their tolerances were huge, the record keeping rudimentary and their attitude was that if it was good enough to liberate Kuwait then it was good enough for the road. Many operations were done by hand and measurements recorded by illiterate mouth-breathers in dungarees.

Our visit was a disaster. When we left my colleague again burst into tears, either at the ineptitude we had seen or perhaps because he'd remembered that the Hummer was Jane's favourite armoured vehicle. The Mishawaka Hummer plant was closed down in 2007.

Random travels

Ireland

In the Nineties the Internet was still a bit of a novelty, great for pics of nudey ladies but hardly the essential business tool it was soon to become. I used to ask customers to email me. Not that anyone did, but I thought it made us look like a high-tech company. I recall visiting a company in Galway who promised to send me an enquiry for our equipment in a few weeks. "*I'll fax you an email*" the man said, confusing me a bit. Sure enough, a few weeks later, I received a faxed email. He'd typed it in Outlook, printed it and faxed it. Four days later I received the same bit of paper by post.

VW's first car factory outside Germany was in Ireland: a plant in Ballsbridge, Dublin, knocking out Beetles from 1950. Ford once had a Sierra factory in Cork. Ireland didn't have much of an automotive industry but I'd fly over and knock on doors anyway.

I met some great Irish engineers, albeit working elsewhere in the world. They all seemed to have that physical characteristic that the greatest real engineers have: something missing. A finger, perhaps a bit of face, or a superfluous chunk of a limb. These guys would put their fingers where others wouldn't.

I worked with one who was brilliant, despite missing a thumb. On long car journeys together he liked to absent-mindedly play air guitar to whatever was on the radio. I wonder what that would have sounded like, played for real, did he make an allowance for the missing digit?

Incidentally my daughter, now grown up, is currently dating a young engineer. He popped round recently with his hand bandaged, smiling; “look, Rich! I’m nearly a real engineer!”

Random travels

Germany

Germany has featured a lot in my career. I have never understood why more British people don't visit. It has so much going for it: Good service, friendly people, decent hotels, reasonable prices, some lovely towns and cities and a mix of beautiful countryside. And racetracks, like the Nürburgring. After my mega-mile Honda went back to the lease company I opted for a slightly sportier company car and ordered a Golf GTI, around the time our business in Germany was really starting to take off. Can you guess where this is going?

I liked the accountant. I appreciated the fact that he'd let me order a Golf GTI as a company car, perhaps in recognition of the good job I was doing for the company. It was silver, had a car-phone fitted, a decent stereo and looked fantastic. Business was good, many of the big German car companies were buying our products, and we were starting to sign up agents and distributors for our products. I was on a roll. We signed an agreement with a big company and, as part of the deal, supplied them with a large stock of measurement instruments at a special rate in order to help them get up and running. We also offered to deliver these instruments free of charge and give them some detailed training. I say we. I mean I. I hope the accountant doesn't read this.

It *was* some years ago and the details are a little hazy now, but I recall that the arrival date of my new Golf GTI and the delivery date of these goods perfectly coincided. As a further co-incidence, with the seats down, all the goods fitted perfectly in the back of my car. I had a Stones Roses CD and a plan.

No-one suspected a thing when I offered to deliver the goods and train the customer personally, and no-one else in the office had seen the recent documentary on telly about the Nürburgring. A place which had me hooked. It is (for those who don't know) a very hairy racetrack that can often be accessed in return for paying a small fee. Technically it is a 'toll road'. It's also the place where F1 ace Niki Lauda burned his face off in a crash in 1976. Even today the air ambulance pops in regularly to scrape up those who got it wrong on the 12.9 mile circuit through the mountains.

The trip over went OK. Customs at Dover were a bit suspicious at a fresh-faced chap with a brand new GTI stuffed full of boxes of hugely expensive equipment, but I drove door to door in a day. It was 666 miles. I stayed overnight and delivered the goods the following morning. The 'detailed training' consisted of me throwing the boxes at a bewildered receptionist and shouting "read the manual" in German. I had to get a wriggle on.

The Nürburgring is only open to the public at certain times. You paid at a little booth, collected a ticket, queued at a barrier and drove as hard as you like. These laps are called *touristenfahrten.* They were, and still are, bedlam.

I have since driven the Nürburgring many times. Actually, since witnessing a near-fatality in 2014, I daren't do it anymore. Back then I was young, eager and thought Niki Lauda was actually born like that. I don't remember much of my first lap.

Those who have been will confirm it's a maelstrom of mentalist motorbikers, badly driven supercars, fast locals who know the lines, practicing racing drivers and assorted maniacs chasing a "10 minute lap" goal. There are stories of cars on the roof with corpses strapped in with laptimers ticking. I just wanted to do a lap.

Timing on such days is actually forbidden so in order to see how close to the magic 10 minute mark I was I played one of my favourite songs – Fools Gold by The Stone Roses, which is a helpful 9 minutes and 53 seconds long. I was just enjoying myself when the carphone rang. It was the accountant, asking how I was getting on with the car. "It's very noisy!" he said. I'm not surprised. I was doing about 120mph on the Döttinger Höhe straight.

I finished my lap, was sick in the carpark (nerves at the realisation at what I'd just done) and decided to get home as quickly as possible. Germany flew by but Belgium was busy and I was short of time to catch the last ferry home. At a place called Jabbeke (where they used to drive land-speed record attempts, ironically) I was stopped by a Belgian traffic Policeman on a motorcycle for driving at 151kph. I had a bit of a meltdown. I had been driving flat out for days with little sleep. And I was a prat. I kicked my car in anger at my stupidity, split my shoe, broke my toe and dented the door. I told the policeman (who must have been laughing behind his visor) that I refused to pay and that I was going to call the British Embassy. He told me I would spend the night in jail. I paid. And I missed the ferry.

Sometimes, when I visit the Nürburgring nowadays, I scan the carpark for lease car stickers and other tell-tale signs that someone is behaving as badly as I did in their own company car. I do hope so. It was cracking fun.

Random travels

Norway

Norway is a fascinating country with serrated skylines of pine trees and mountains, all rather monochrome and melancholy but beautiful too. For one memorable meeting there I took two flights, the second being a bumpy ride over the mountains in a Twin-Otter propeller plane to land on a runway seemingly the length of tennis court. The sea was frozen. A taxi ride up into the snowy mountains took me to a truck manufacturer.

These guys had a great reputation for quality yet everything I tested was wrong. In fact, such was my enthusiasm for helping them, and their gratitude for getting some good test data, that we inspected the whole place as part of my product demo. I realised, at the end of a long day, that they now had no need to buy anything as I'd done their job for them, reports and all.

They started work early and finished early too. I was astounded to find people driving these huge dumper trucks home, rumbling over the scenery and vaguely following the course of the narrow, snow covered lanes. I took a taxi back to the airport and found the only Scandinavian who didn't speak English. He was a very old chap in a very old Mercedes. He spoke German and I asked how long he'd been working there. He was a German soldier who, in 1945, decided he would rather stay in Norway than go back to a broken Germany.

I got to the airport really early to find there was only one flight a day out of there. It was too cold to hang around outside and nothing to do there anyway. I recall sitting for six hours, watching the clock tick down to my flight.

Random travels

Poland

Poland was always interesting. I had a Polish colleague who did the hard work and after a while I noticed a pattern to our meetings. I'd speak English and my colleague would translate to the QA manager of the sites we visited. If he agreed with our proposal he'd call the production manager who usually had the authority to sign an order.

Without fail the factory managers always had a *huge* moustache. At one place in Warsaw the manager turned up and was clean shaven. I told my translator to tell him that someone so clean shaved couldn't possibly be management, and to bring the real boss. He nervously translated and the man left the meeting, chuntering in Polski, and came back with a man with a 'tache the size of sofa.

After this we went for a few beers in Warsaw and I got mugged in an alley (serves me right for being so cocky earlier). To cheer me up my colleague took us to see Warsaw's famous palm tree. It was just a miserable tree stump.

Random travels

Russia

I went to Moscow to see a customer who regularly paid us $250k for $50k's worth of someone else's product but refused to tell us what he was doing with it. My boss was a bit uncomfortable so sent me to investigate. An easy trip, I thought, and another interesting stamp in my passport. The Russian embassy in London made life as difficult as they possibly could, took a lot of money, and finally gave me the rubber stamp to travel.

The 'company' was a university professor in Moscow who ran a business on the side, using his brightest (and prettiest) students as staff. He insisted we met in a restaurant where he flatly refused to talk about any business related matters and talked, loudly, about Hitler, Jews and Americans. His views made Vladimir Putin seem like Jeremy Corbyn. I started off politely, asking about his business, hoping to turn things around to what he was doing with a container full of complex equipment, but he just kept filling my glass with vodka, my plate with potatoes and my head with extremism. I gave up.

Moscow was an incredible place, but scary. I stayed at the Sputnik Hotel, chosen because I liked the name. It was an old military building converted to be a tourist hotel. The miserable receptionist looked at me like the filthy capitalist pigdog I am when checking in. The lift was out of order and she shouted at me for walking on the carpet "*NYET!*" and made me walk around the edge of the room so as not to wear it out. The room stank and was filthy. It was on the 8th floor and had a balcony. The balcony door was held shut by a bent nail and when I stepped onto the balcony I saw the handrail was fitted not at hand-height but at knee-height.

I went out for a walk. The receptionist shouted at me for walking on the carpet, then shouted "*KLUJ!*" which is Russian for key. I don't like leaving my key at reception in dodgy hotels as they know your passport, money and belongings are then unattended in your room. I told her "*nyet*" and tried to walk out but she locked the door remotely. We then took it in turns to shouts "*KLUJ*" and "*NYET*" at each other for a bit before she switched to English "*YOU! GIVE ME KEY! NOW!*" to which I told her to open the bloody door and let me out. Then she called security and I happily gave my key to an ape in a suit with a car-crash face, before finally being allowed out.

On returning to the room (treading carefully around the precious carpet) I tried to sleep but thought I could see someone on the gloomy balcony outside, trying to get in. I switched the light on and the shadow went away, then heard a distant thud, like a car door being closed. I didn't sleep. In the morning I had breakfast which was boiled egg and slices of mystery meat and dropped my key at the unattended reception. Outside I went looking for a taxi. Half of the carpark was taped off with lots of Police about, and a Lada with a stoved in roof, beneath my balcony.

On the flight home I got chatting to a British chap who was a regular in Russia. He told me that he once sat on an Aeroflot flight next to a Russian with a goat in his lap. On objecting to being piddled on by the goat he was told by it's owner that it wasn't a goat, but a bag, pointing out the rope tied from it's neck to it's tail, slung over his shoulder.

Random travels.

China

My travels had started (many years before) in South Wales and soon took me around the UK, continental Europe and occasionally the US. I had one very memorable trip to China to visit a manufacturer of electrical components for the automotive industry. It was a long flight to Hong Kong where the hotel sorted me a visa for China, then a helpful person put me on a ferry down the coast to China. It was kind of them to help, but it was the wrong ferry.

After a few hours I had the feeling that I was heading in the wrong direction. No-one spoke English. I had an address on a piece of paper in English and Chinese. I may as well have been showing people a piece of used toilet paper. The name of the port I should have disembarked at never came up and like a drunk on a bus, I was the last man aboard when it reached its final destination.

Up the coast (or down, I never found out) someone from the company was waiting to meet me. But not here. I left the ferry and was instantly bombarded by clamouring locals who had never seen a lobster-faced Englishman in a suit and tie before. Eventually a single white face passed in the crowd and spotted my distress. He too was English, there to check on a shoe factory he owned. He had a Midlands accent, like me. When I told him the name of the small village I come from he thought I was joking. I showed him my passport. He was from the same place! I shocked him further when he said which house he lived in and I recalled the private plate of his wife's perpetually badly parked BMW. An incredible co-incidence. We had a chat and parted, I jumped in a taxi, handing the driver the address.

My logic was that if I had travelled three hours in the wrong direction on a ferry, and a taxi travelled at roughly the same speed as the car, I was at worst six hours from my destination. We soon left the port and were on a badly surfaced road through the countryside. The driver spoke not one word of English. The car shook violently at speed. There was no instrument binnacle. I had a bad feeling. My mobile rang. "*Where are you?*" I had no idea. "*Whatever you do – DO NOT GET IN A TAXI*". Great.

I watched the time. I had decided at the six hour mark to jump out, regardless of location. We seemed to be in endless countryside. At five minutes to the six hours I was relieved to see a big hotel on the edge of an industrial estate. We pulled up to the gates and the guard put a gun to the drivers head and they both started shouting at each other. I practically gave birth on the back seat. Well, my waters broke. I later learned that it was an illegal taxi, and the fare I'd paid was many times the going rate.

The factory was nearby. Western owned, well established and very efficient. Workers got just $1 an hour there so there was little need for full automation. Old machines cut, pressed, screwed and coated parts and workers moved them around the huge factory by hand. The machines were reliable but old, made in America and Europe.

When machines broke the Chinese workers would try and fix it themselves. They wouldn't take a spare part from stores, they'd try and repair it with old bolts, nails and even bits of litter. China has a terrible reputation for copying Western products, regardless of patent or copyright, yet these factory workers were incredibly innovative and resourceful in making repairs to their old machines. If only they could create instead of just copy and bodge! There was no chance of helping these workers develop their careers though; they earned enough in six months at this factory to travel inland and buy a farm so staff turnover was high.

The few Western guys working there were friendly. They all had failed marriages, bitter tales to tell about the west and spent their free time drinking heavily and indulging themselves with young Chinese girlfriends. It was a strange world they inhabited.

The factory was on a hillside, protected by a big fence. On the other side of the fence, at the back of the site, was scrubland. There, a heavily pregnant woman lived in a shack made from wooden pallets, covered with a tarpaulin. She sat feet away from me, in the dirt, and we made eye contact. I couldn't get my head around the colossal difference in our lives. How come I was so lucky? I later heard the cries of a newborn. This really hit me. It's one thing laughing at a BL bodge in the Midlands, but on the other side of the world, missing my wife and own baby daughters, the fun had gone.

I had a farewell meal in a local restaurant of 'pig hand' and 'salted ox penis broth' and too many Tsingtao beers, and again bumped into my neighbour at the airport on the way home. Salted ox penis broth tasted rather like Cowley canteen's ham sandwiches, actually.

After this trip my career took a different path, and I left behind the world of Quality Control, inspection and factories to do something quite different. But there was one little piece of unfinished business, which I would return to many, many years later; Longbridge.

Longbridge, and a scientific experiment

The very name, Longbridge, seems to conjure up images of massed industrial action. We've all heard tall stories such as cars being built with brake discs on one side and brake drums the other. I read somewhere that they supposedly lost half a dozen cars a week due to theft - people somehow just drove them off site. In one infamous strike in their darkest days £10 million pounds worth of productivity was lost due to a dispute over staggering tea break times.

Sadly I never had any dealings with Longbridge myself, other than to once work with an oik of an ex-QA Engineer whose party piece was to drop his trousers, stuff his genitals between his legs, bend over and shout "*fruitbowl!*" I was told he was one of their best guys. MG today, sometimes finish off Chinese built cars there but it's a skeleton staff compared to the 25,000 that once worked there.

I really wanted to see if Longbridge's products were as bad as many would have you believe - it had long gnawed away at me. Now working in another industry I no longer had access to the measurement equipment needed for an accurate assessment of their build quality, and of course MG Rover hadn't built a new car in a decade.

In 2014 I colluded with a few fellow motoring masochists and came up with a plan. We would buy examples of 'industrial unrest Rovers' and put them to the test. Car manufacturers use rallying to demonstrate the durability of their cars and Formula 1 is considered the technical pinnacle of motorsport. For reasons that made perfect sense in the pub we agreed to combine the Paris-Dakar rally and the 1958 Casablanca Grand Prix circuit with a roadtrip to Morocco.

I bought and collected a Rover Metro from a man in Nottingham. It was £299. The seller waved me off with a friendly warning: "*You do know these are sh*t, don't you?*" I wanted to find out for myself.

Setting off at half past bastard o'clock in the morning my co-driver Darryl and I ragged down to Portsmouth in the Metro to meet fellow scientists Baz and Jimbo in their Maestro and Ben and Chas in their Rover 416; we boarded the ferry to Spain, and spent 24 hours clinging to the bar trying to put our paperwork in order. The nautical shortcut across Bay of Biscay was stormy yet I wasn't bothered that I hadn't put the handbrake on the Metro - these cars were not coming back.

I have an almost religious approach to prepping cars for roadtrips – if you replace just one rusty washer, you have to replace the nut, and then whatever it's holding … *ad infinitum* until you've done a full resto job. Changing one scabby connector upsets old wires and creaking circuits, setting off an endless chain of catastrophic maladies on a worthless car. I call it mechanical karma. Leave it all alone. They've managed 20+ years and only need last a thousand miles more.

I wanted to leave them as factory standard as possible, to see how well they were screwed together by the bolshie Brummies I'd heard so much about. We spent the maintenance budget on pith hats and stickers.

Spain was the easy part of the trip. 24 hours of motorway to get down to the port in Algeciras for the crossing to Africa, a steady drive for the cars. Yet at the first hill coming out of Santander the Maestro overheated and broke down. It was the first time I had chance to look at the thing.

When Rover stopped making the Maestro in Cowley they sold unused parts to a company in Bulgaria who bodged a few together. Unsurprisingly, they didn't make as many as planned and sold leftover bits back to a company in Ledbury, England, who assembled the car and sold it buyers in the UK who were so keen to buy British they'd accept an ancient design with a KPH speedo, wrong-way wipers, Ox-cart comfort and a 1275cc engine first built in 1951. They are, supposedly, rare and collectable now.

The overheating was traced to coolant that had never been changed so we emptied the slurry from the rad, topped it up and went on our way. Driver Baz normally enjoys a Merc C63 so when he screamed over the radio "*slow down! It's knackered! It has no power!*" no-one had any sympathy, not even startled Spanish traffic cops on the same frequency as our walkie talkies. *¡Hola, amigos!*

We arrived very late for the departure of the ferry to Morocco. Fortunately the ferry was very *very* late and we made it in the nick of time. We waved goodbye to Europe and crossed the Strait of Gibraltar, arriving in Tangier, a few miles from Europe geographically but as far removed from civilisation as an overheating Maestro is from being stylish intercontinental transport.

Angry men in ill-fitting uniforms with guns shouted at us in French while locals hassled us. To get into Morocco you have to complete a form to import yourself, another form to import your car, buy insurance for the car, then buy local currency which we called "stinkies" on account of their much-handled aroma. Much of the paperwork is in foreign (Arabic, anyone?) it's hot, late, and there's a jostling scrum of aggressively friendly locals clawing at your passport and wallet. Still, the cars had held up and we'd made it to Morocco.

We headed south for an overnight stay in what the Internet had assured us was a clean, safe and friendly place just south of Tangier called Asilah. The Internet was wrong.

The hotel owner spoke some English and we got a four bed, two-roomed apartment for £30. He was quite friendly until it came to being seen with us in public. He poked his head out the door to first ensure it was safe before directing us to a nearby cat-infested restaurant for tagines, chips and olives. The cars were locked in a compound, next to a dead Renault 5 and a British registered Mercedes G-wagon that looked utterly kaput.

Leaving Asilah in the morning we aimed to avoid the coastal motorway and give the cars a ragging on the old road that ran parallel to it, rally style. These roads are potholed, have crazy subsidence in places and the rain teemed down. We got lost. The Maestro conked out in an open sewer and as Baz was attempting to google "Rover specialist near أصيلة القرف الشارع" it miraculously sprang back to life. Thank you, whoever made this car way back when, thank you.

Traffic is sparse on the open road but there is an incredible jumble of transport to be avoided in the towns. Motorbikes converted to pick-up trucks seemed popular, as well as Renault 4s and 5s, Peugeot 205s which also serve as taxis and smoky old Merc saloons with up to half a dozen people rammed in the back, all spewing pollution you could collect in a jar and sell as Moroccan Marmite. Our cars, though, were holding up surprisingly well.

I'm not sure what I liked most about the Metro; It's incredible fuel economy, the boingy handling or the fact that we could drive it across Africa without worrying about the inevitable scrapes and scuffs suffered along the way. The hydragas was saggy and the brake warning light was on, but "they all do that, Sir". Hydragas is a brilliant invention and would work wonderfully on modern small cars on Britain's traffic-calmed, potholed roads, given another chance.

Owning a worthless Rover is fun. Ramming another worthless Rover at every opportunity in heavy traffic more so. The 416's back bumper took a dozen shunts to dislodge, but it isn't a Rover in the darkest, Longbridgist, Leylandist sense of the word. It's a leftover from Rover's collaboration with Honda. Japanese underpinnings with added wood cappings, velour seats and an automatic gearbox, first owned by a Doctor who paid more for the nightfire red paint option when new than Ben and Chas paid for the complete car. £180 gets you a reliable and comfortable car that, apart from vandalism by our fellow 'scientists' and the word TW*T inexplicably embroidered on a headrest by a previous owner, worked perfectly.

Aside from the overheating A-series engine in the Maestro all the cars were fine, if a bit battered, by the time we reached Casablanca. Our test hadn't quite covered the Paris-Dakar route but we'd done a good enough chunk of it to know that our Longbridge built Metro was perfectly reliable transport.

Stirling Moss won the Casablanca grand prix in 1958, a race organised by the King of Morocco to showcase his country; it's a blast down the coast road south of the city centre then a rough left, left, left, left square back to the start. No faded colonial beauty here, nor Monaguesque opulence, as the site of the original pit lane is now a KFC.

The city is overcrowded and chaotic. Traffic is horrific. A blinkered population of five million on roads clogged with carts pulled by emaciated mules, overloaded lorries, mopeds and old French hatchbacks. We also saw a Rover 820 and a 618 and felt strangely envious of their luxury.

We dumped the cars and took a taxi to dinner, whereupon our driver ran over a couple attempting to cross the road, backed up to get them from under his front wheels, then eventually dropped us at Rick's Café. Of all the gin joints in all the world, why did we walk into this one? Rick's is brilliant. It's a bar and restaurant carbon-copied from the classic film *Casablanca*, run by an American ex-diplomat who serves the most fantastic food, cold beer and obligatory G&Ts. The décor is colonial splendor with Arab touches and a Sam on piano. It's beautiful and a complete contrast to the hell-hole just outside the door. I was hoping for the company of Ingrid Bergmann but got Rover owning Brits in flat caps and pith helmets enthusing about their cars.

The most BL of the lot was the Maestro. I expect that some readers may be unhappy that Darryl smashed in its doors and bonnet with a golf club when it overheated for the sixth time but these cars, as rare as they now are, are worthless. If it was so special why could we pick one up for the price of a few drinks at Rick's? We could only blame its unreliability on poor maintenance and the fact that Baz had zero mechanical sympathy. It certainly held together well. I think Jimbo had bonded with it but like a holiday romance it had to end.

Ben and Chas had played a kind of BL Buckaroo with their 416. They claimed it was a reverse engineering project. They removed as many Rover parts as possible as they went along, arriving in Casa without most of it's fake wood trim and superfluous luxury components, leaving mostly the Honda it was built on. But it got there, no fuss.

The Metro had just 41k on the clock and had been well maintained by a Nottingham pensioner, but watching a wing burst open like a bit of wet cardboard after a little collision with [*removed in case their insurer is reading this*] we decided that it needed about 200 Kg of strengthening to make it safe, which would cripple the lightweight, chuckable feel that made it so much fun in the first place.

These cars might be a bit crap judged by modern standards but we loved them and I'd like to think that the men of Longbridge who originally built them would approve of our experiment. And the results. After mistaking a mosque for a disco (I *thought* the music was rather peculiar) we retreated to the safety of the 14th floor of the Ibis, watched the chaos on the roads below and planned our retreat.

Cars cannot be left in Morocco. Their import is tied to your own entry paperwork and passport so they have to be exported and disposed of elsewhere. We trundled 250 miles to Tangier, past fields of miserable camels and gawping locals and caught a late ferry back to Spain which was overloaded and listing at a good 20 degrees when backing out into open water; a last bit of excitement from Africa.

We had a flight the following day but how to dispose of the cars, vaguely legally? They were worth less than a ferry ticket before we left; in their battered state they're not worth the time or money to repatriate. We had tried to donate them to Oxfam. Oxfam didn't want them.

The Maestro was left outside the gates of a Spanish scrapyard with the keys in it and the chaps piled into the other two cars. The Longbridge built 416 and Metro crossed into Gibraltar where we fed the car's UK logbooks to the apes. We had tea and cake at a Café and paid the elated waiter with the keys for both the Metro and 416 parked outside before jumping aboard a budget flight back home to cold, wet Brum.

Africa had been insane. We had only dipped our toe into the Dark Continent and each left two stone lighter with empty wallets and frayed nerves after just a few exciting days. I can think of no better destination for a Rover. There's a wonky Metro hubcap on my office wall and I'm only sad I didn't repatriate the rest of it.

So, the result of our experiment: they're all winners, these cars. They did a trip that was considered beyond them by so many, stood up to the worst abuse we could chuck at them with zero maintenance or preparation and made us laugh. They have a battered, buggered and faded appeal that I've learned to appreciate. This experiment left me with a respect for Rover that I never had before.

If only we'd better appreciated these cars when new then perhaps we'd still have a volume British car manufacturer of note, despite the sometimes slapdash (but often amusing) antics in quality control.

The end.

About the author / thanks

Rich Duisberg is an occasional writer and presenter who once earned a living helping companies to put new cars together properly, but who now spends his free time taking old cars apart and losing the important bits. He was once threatened with legal action by very old DJ Timothy Westwood and Le Mans champion Derek Bell refused to get in a Morgan 3 Wheeler with him.

With thanks to the owners and employees of the aforeunmentioned great little company who kindly employed me. Thanks for the love, my amazing ladybabies: Freya, Hattie and Claudia, to Johnny and Babs for pointing me in the right direction, Paul Nixon for the cover artwork, Dr O and David Chapman for the prüfereåding, Reg 'underscore' Local, Richard 'SniffPetrol' Porter for the foreplay request and cover quote, Michael Downing for his encouragement, Bernie Nyman for the legal work, random internet forum types for their feedback, those who helped get this off the ground by pre-ordering on KickStarter and ***a huge and sincere thanks to you, reader, for buying my little book.***

Rich Duisberg

Email – Rich@MotorPunk.co.uk
Twitter - @TheDuisbergKid
Web – www.MotorPunk.co.uk

PS – I'm now working on something with the working title; 'The Bumper Book of Roadtrips'.

Book created in a slightly arse-about-face format using Createspace for Amazon. ISBN-13: 978-1532719790 (CreateSpace). ISBN-10: 1532719795.

Printed in Great Britain
by Amazon